# PACIFIC MEMORIES

## WAR AND PEACE IN FAR AWAY PLACES

RICHARD S. PARKER

# Publisher's Information

EBookBakery Books

Author contact: margieflanders@aol.com

ISBN 978-1-953080-17-2

© 2021 by Richard S. Parker

ALL RIGHTS RESERVED

No part of this work covered by the copyright herein may be reproduced, transmitted, stored, or used in any form or by any means graphic, electronic, or mechanical, including but not limited to photocopying, scanning, digitizing, taping, Web distribution, information networks, or information storage and retrieval systems, except as permitted by Section 107 or 108 of the 1976 United States Copyright Act, without the prior written permission of the author.

# Acknowledgments

I want to thank all those who have given me help and support through my "third career" as a writer, for all the op-ed pieces, and my now second book. They include my family, my son Eric Parker and cousin Betsy Palazzetti, and cousins Don Hall, Ann Taylor, and Karl Parker.

Those who have championed my work include my dear friend Archie Thornton from the "Mad Man" days, my friend Sandy McCaw and my neighbor, Eleanor McSally. Special thanks to my editors, David and Marguerite Flanders, who have hung in with me for the last four years, and to I. Michael Grossman, for producing what you are now reading here. I'm thankful to you all.

## Dedication

*For my grandfather, Dr. John S. Newcomb, who went to sea after his young wife died. He served as ship's doctor on an ocean liner that sailed from Vancouver to the Far East. Our family went with him in spirit. Upon his return, we learned a lot from his inside stories concerning the "mysterious Orient." These stories kindled in me a lingering curiosity that fueled my adventures during the Occupation of Japan.*

# Contents

**Preface** ............................................................................................ **viii**
**Part One: Basic Training** .................................................................. **x**
   Early Memories of the Pacific War ................................................ 1
   Why I Enlisted in the Infantry ....................................................... 5
   My Career as a Sniper .................................................................... 7
   My Home for Three-and-a-Half Years: ....................................... 10
   Camp Breckenridge ..................................................................... 12
   Infantry Training ......................................................................... 14
   An Infamous Bivouac .................................................................. 18
   I Become the Poison Gas Non-com ............................................. 19
   The 98th and Rommel's Africa Korps .......................................... 22
   Yoo Hoo Ben Lear ....................................................................... 24
**Part Two: Tennessee Maneuvers** .................................................... **30**
   Tennessee Maneuvers .................................................................. 31
   The Road to Possum Hollow ....................................................... 33
   Country People ............................................................................ 37
   Fort Rucker ................................................................................. 40
   Working with the Artillery .......................................................... 46
   In the Deep South, a Statue to a Bug .......................................... 47
**Part Three: Trip to Hawaii** ............................................................. **50**
   Trip to Hawaii ............................................................................. 51
   Life on Kauai .............................................................................. 53
   Special Duty in Paradise .............................................................. 55
   Island Entertainment ................................................................... 58
   Training on a Wet Mountain ....................................................... 60
   Na Pali Valley .............................................................................. 61
**Memories of the War and Japan: Sketches by the Author** ............. **66**
**Part Four: Oahu** ............................................................................. **68**
   Back Country Oahu .................................................................... 69
   Fort DeRussy .............................................................................. 71
   From Waikiki to Tent City .......................................................... 72
   Navy Special Duty, Oahu ............................................................ 74
   On the Positive Side .................................................................... 79
   Invasion Tactics ........................................................................... 82
   Lust and Leprosy in the South Pacific ......................................... 84
   Friendly Fire ................................................................................ 89
   Hotel Street, Honolulu ................................................................ 93
   The Atom Bomb and the End of the War ................................... 96

**Part Five: Occupation of Japan** .................................................................. 102
    On Our Way to Japan ................................................................................ 103
    Landing in Imperial Japan ........................................................................ 108
    Amphibious Vehicles ................................................................................. 112
    On to Osaka .............................................................................................. 115
    Saving Prisoners of War ........................................................................... 118
    Hunger in Japan ....................................................................................... 120
    Benjo, the Very Strange Japanese Toilet ................................................. 122
    Cabarets (Night Clubs) ............................................................................. 124
    Bob Wulfhorst's Kamikaze Interlude ...................................................... 125
    The Saving of the Shrine City: Kyoto ..................................................... 126
    Visiting a French Priest in Occupied Japan ............................................ 130
    Sayonara, Japan ........................................................................................ 134

**Other Recollections** ......................................................................................... 138
    Chaos in the Philippines .......................................................................... 139
    Midway: The Epic Battle of the Pacific War ........................................... 141
    The Submachine Gun .............................................................................. 144
    An American Flying Boat Lands in Japan .............................................. 144
    General Douglas MacArthur: .................................................................. 146
    Sex Life in Occupied Japan ..................................................................... 149
    Dwellings in Imperial Japan .................................................................... 150
    Religion in Imperial Japan ...................................................................... 151
    The Japanese Zero ................................................................................... 153

**Postscript** ........................................................................................................... 155

**About the Author** ............................................................................................. 157

# Preface

Memories can be fragile, but my memories of the Pacific war, and the Occupation of Japan, remain strong and robust. This is because they are remembrances of events that took place that were so unusual, so wild and weird, so unlike anything that I would have encountered in civilian life, that my memories of these colorful events persist, although the details of what happened can be sometimes lacking.

Many of the events I discuss were secondhand when stories were told, and some were even third- or fourth-hand events, remembered from bull sessions and other military gab fests that took place. There are often times, particularly when one is at sea, when the wait of the hurry-up-and-wait syndrome takes over. At that time, the emptiness of 'wait' needs to be filled, and so, stories that I heard in these sessions, I think, were sometimes invented just to fill that empty time. So, I cannot vouch for the veracity of my Pacific Memories, but they were, to me, interesting examples of what could have happened, and sometimes did happen, so many years ago.

# BASIC TRAINING

## Early Memories of the Pacific War

WHILE STILL A CIVILIAN, I became part of the war effort. When the bombing of Pearl Harbor in 1941 brought about the entry of America into World War II, I was in my junior year at Rhode Island School of Design. For the rest of the year, it was difficult to concentrate on school subjects with climactic battles raging in Europe and the South Pacific. When school ended for the year, I still had not received the 'Greetings from the President' that would summon me to service; so, I decided to get a defense job, at least for the summer, or until I was called up.

When President Roosevelt demanded what seemed to be impossible quantities of planes, tanks and other needed war materiel, some said that we did not have the industrial capacity to achieve these goals. They were right. After the many shutdowns forced by the 1929 Depression, America did not have the ability to produce the necessary military equipment - not only for our war effort, but also for our wartime allies, England and Russia.

However, in spite of almost impossible-to-solve difficulties, America did reach Roosevelt's goals by harnessing all aspects of America's industrial potential. Almost everyone in America was involved in the war effort in some way or another, whether it was a family member in the military, going along with the rationing of food or gasoline, having a victory garden, or even just giving up nylon stockings. This involvement fostered an incredible spirit of cooperation very much in line with the concept of subcontracting, one of the most important components in reaching war production goals.

The 'big boys': Ford, General Motors, Boeing, Kaiser, and other major industrial producers were able to shift their production almost entirely to the war effort. Through subcontracting, practically every other manufacturer, no matter its size, could become an important part of the effort. As an integral part of military manufacturing, subcontracting was of immense help in meeting war production goals, and incidentally allowing the continued production of some items still needed for the civilian world.

While much military equipment was too large or too complicated for one company to handle alone, a group of manufacturers, under the direction of the War Office, and working from blueprints held in common, would subcontract production of specific parts that they did not have the equipment to handle. These manufacturers, in constant communication with other members of their team and the War Office, made sure that each of these different parts was constructed properly, and fit exactly with parts from other manufacturers, before shipping these parts to another subcontractor for further work, or assembled and sent to their final destination. In a short time, subcontracting became America's secret weapon, allowing the creation and accomplishment of war production schedules that otherwise would never have gotten off the drawing board.

Looking around for a suitable employment opportunity, I heard that the Pantex Pressing Machine Company, in Central Falls, Rhode Island, was a good example of a company that, while still making their usual pre-war products, was now active in war work. Because of the wartime shortage of workers, it was easy to get a job at Pantex, a leading maker of equipment for dry-cleaning and laundry establishments, and still making pressing machines for Army and Navy post exchanges. In those war years, however, the real work of Pantex was subcontracting the production of many items that had nothing to do with dry-cleaning but were essential to the war effort.

My job at Pantex carried the fancy title of expeditor; but in fact, I really was just a truck driver. What the title of expeditor really meant was that I would drive a pickup truck, loaded with war materiel that Pantex had partially completed, to another factory, where more work was done on it. If additional work was needed after that, I would cart it back to Pantex, or possibly off to another subcontractor. This went on until the project was put into final assembly and sent off to the military.

That is how it worked with a major defense weapon, the Oerlikon Gazda rapid-fire cannon.

In Germany, toward the end of WW1, Reinhold Becker began to manufacture a weapon called the Becker rapid-fire cannon. When the post-war Versailles Treaty banned further production in Germany of this and other weapons, manufacturing of the Becker was transferred to Switzerland. A factory was created for it in a district in the northern part of Zurich, called Oerlikon. Eventually the gun became named for the district where it was manufactured.

Just before World War II, the Oerlikon, known for its naval applications because it worked well at sea, rarely broke down, and, having effective range and adequate firepower, was on order by England. But, in spite of entreaties by Lord Mountbatten, Britain was slow to implement their order, and only a few were shipped before the war began. After hostilities started, Switzerland, now surrounded by Nazi forces, could not make further shipments to England. As it was the best weapon of its kind, and the Germans knew about it, there was concern that plans for the Oerlikon might be acquired by the Nazis.

The international representative of the factory where the gun was produced, Antoine Gazda, an anti-Nazi Swiss citizen, had overseen the modification and improvement of the Oerlikon, now a 20-mm rapid-fire cannon. Deciding to keep it from falling into the hands of the Nazis, he contrived to bring the plans for this weapon to America. The American military, when first approached by Gazda, had turned him down, as they were accustomed to working with guns that fired 50-caliber bullets. However, Rhode Island's governor at that time, William Vanderbilt, hearing of Gazda's weapon, sent an emissary to New York to encourage Gazda to come to Rhode Island. His offer was based on the availability of the many machine shops in the state, ready and willing to manufacture the Oerlikon. They promised to do this through subcontracting the various parts, a move quickly approved by the American War Review Board. Establishing a headquarters in Providence, the Gazda Gun Command Center took on the responsibility for allocating the production of the weapon to various and diverse companies.

Pantex, already deep into subcontracting, was assigned, among other things, the responsibility of producing a rather complex gun sight for the Oerlikon, making use of their expertise in working with metal. The sight consisted of a series of concentric wire circles joined by radiating bars, with a tab on the bottom where it was clamped to the gun. The special grinding on the tab, needed to make the sight fit properly, could not be done by Pantex. So, several times a week, I would bring large crates of gun sights over to East Providence, home of giant Phillipsdale Wireworks. They would carefully finish the work on the tab - a small job, but necessary to make the sight fit properly. The next step would be for me to pick up the gun sights that had been treated by Phillipsdale and bring them back to Pantex, to be added to other parts of the gun that were being worked on.

With Antoine Gazda supervising, the work went well, and in a short time, many guns were in active use, effectively altering the balance in submarine warfare to favor the Allies. The Oerlikon Gazda was also successful for anti-aircraft protection, and over 23,000 of these were employed by the navies of both England and America.

They were also in use by the U.S. Merchant Marine. Early in the war, Nazi submarines, rather than using expensive torpedoes, would often surface to shell cargo ships. Now they could be driven off, or even sunk, by fusillades from the Oerlikons. One of the best weapons of World War II, the Oerlikon Gazda was so important to the war effort that Antoine Gazda was assigned Military Police protection, both at his office in Providence, and at his home in Narragansett.

There were other subcontracting assignments, as well. About once a week, with some help, a large torpedo warhead which Pantex had already worked on, but which needed further machining, was loaded on to my pickup truck. From Central Falls, I would drive over to Fairlawn, a working-class section of Pawtucket, to deliver the warhead to its next subcontracting operation. When I arrived at my destination, instead of entering the formidable gates of a giant defense plant, I would alight from my truck, go to the door of a three-decker wooden tenement, ring the bell and be greeted by our subcontractor, an elderly man with a strong English accent. After exchanging pleasantries, we would walk over to my vehicle, and between us, get the warhead down off the truck; roll it across the courtyard to the cellar door, open the door, and together, an old man and a young one, would carefully slide the warhead down the cellar stairs and roll it across the floor to his lathe.

Machinists usually love their work, and often have considerable equipment in their basements where they can work on their hobbies when not working at their jobs in the factory. Like many of his kind, our subcontractor, a retired machinist, had a well-equipped workshop where he could do a precise and difficult grinding operation on the warhead. At the lathe, we worked together, lifting off the warhead that had been worked on, boosting up the new one, and locking it in place. We would then roll the completed warhead across the floor, up the stairs, across the courtyard, and together, hoist it up onto my truck. And that's an example of how Americans of all ages, working together, won the war.

## Why I Enlisted in the Infantry

AFTER PEARL HARBOR, RHODE Island School of Design began to offer a course in camouflage. As it seemed a logical progression for me, I signed up, and found the study of the art of camouflage quite interesting. The instructors even took us up in small planes to see how things looked from the air, and we studied ways of protecting buildings and equipment from enemy observation. From this start, I knew it was what I wanted to do. My course was clear: at the end of the school year, my junior year at RISD, I would sign up for the Army Engineers and apply for the camouflage section. This would be simple to do, as I already had the documents for this application

Alas, my teenage pride led me to a not so wonderful change of plans. I was walking down my street with my mother, when we met a neighbor, a somewhat self-important person. After a perfunctory greeting to my mother, she turned to me. "Oh," she said, "your mother tells me you are going into camouflage. That is wonderful for your mother. You will be safe, and she will not have to worry about you." She paused for a moment, letting this insult to my teenage pride sink in. *Safe for my mother indeed*, I thought, and my blood began to boil. Then she finished me off by announcing triumphantly, "My son is going to be a flyer!"

The next day, I went to the recruiting office and joined the infantry. It was, I thought, just what I was looking for: a place where no one would accuse me of being "safe for my mother." I was soon to find out that the infantry was a lot more than that.

The training that followed was rugged and exhausting, with not very good food or living conditions. Wallowing around in mud and slush, sleeping on the hard, cold ground, and marching endless miles with heavy packs was difficult physically; but the guys in the outfit were mostly salt-of-the-earth types, good to pal with, and to share the tough life of the infantry. They did not offer much more than that, which was plenty enough. However, I did survive the war, but my neighbor's son did not. His plane crashed while involved in a training exercise and he did not survive.

On the other hand, from what I have read, the camouflage section of the engineers was filled with artists, many of them top-rated in their careers; and were the people who knew their way around the way things worked in the art world. Some of the young men who went into camouflage made connections that were most helpful after the war; for instance, Ellsworth Kelley, who became a star of the abstract expressionist movement.

Occasionally, I tried to do some artwork, but painting equipment was difficult to fit into a full field backpack, and time and place to paint was not often available. However, I did make a few sketches of fellow soldiers, and a few tropical landscapes, but I usually did not have the art supplies needed and had to draw or paint on whatever was available, sometimes even using old envelopes in place of art paper. For many years, these sketches sat in a box marked "Army." Recently, looking for something else, I unearthed them, and found that they brought back memories. Each sketch reminded me of a landscape or pal, or a story of a time when a short conversation with a neighbor changed my destiny.

## My Career as a Sniper

AFTER ENLISTING, I WAS sent to Fort Devens, a reception center where, at all hours, we were tested, outfitted, and assembled for our next destination. Not a very elite-appearing group, my companions seemed to be the scrapings of the bottom of the draft barrel. Groggy from the often nocturnal activity, I followed others onto an elderly steam train that would take us to where our training was to take place, Camp Breckenridge, Kentucky.

On our way, we were often shunted off to sidings, as freight trains, obviously loaded with war materiel, roared past. Crammed for two days and two nights into ancient Boston and Maine coaches that sported antique chandeliers and equally ancient toilets, it was traveling back in time. The seats were pulled, backs together, to create a unit seating four, so that two of us sat facing a second pair, knees together, for two grueling days and mostly sleepless nights. The duffel bags holding our possessions were crammed into the space provided by the back-to-back seats. Meals, provided in a baggage car, cooked with oil stoves set in a bed of sand, were served from so-called G.I. cans (looking remarkably like garbage cans). The food was what you might expect under those conditions. It was mostly stews, with the dessert, usually peaches or other canned fruit, dumped on top of the stew. After eating, there did not seem to be a way to wash our mess kits. However, after our first meal, word was sent around that before our next meal we should douse our mess kits in boiling water. The next day, arriving at the mess car, I spotted a GI can filled with dirty-looking water. "Is this where I dunk my mess kit," I asked? "Hell no," was the response, "that's the coffee!" On the third depressing day, I arrived in the pouring rain at Breckenridge, a camp as new and unfinished as I was. The grim surroundings did not help. An ironic morale builder, a sound truck, blared military music as I slogged through muddy grounds that looked like the end of the earth, to our assigned barracks. Being really beat, I was hoping for some time off; but for some of us, that did not happen. Without time to rest and recuperate, those in a small group were assigned special duties, while the rest of the new soldiers - many of them seemingly not in much worse condition - were given time off.

My assignment turned out to be mind-boggling. For some unfathomable reason, Buck Private Parker became the Provo (provisional) Sergeant of the Guard, given a 45- caliber pistol and was told that I was in charge of the security of the sprawling grounds constituting Camp Breckenridge. This involved supervising a twenty-four-hour guard duty session for the Camp. To me, it was a fraught situation, as I had never had a gun in my hand and didn't even know how to insert the bullet clip. To make it all the more interesting, a seriously deranged escapee from the psycho ward was thought to be somewhere in the camp area. Fortunately, he did not turn up while I was in charge; and somehow, I managed to set up the next three shifts and sent them out, hoping for the best. Thankfully, nothing untoward happened; but I was without much sleep in the 24-hour guard duty period I was responsible for. Needless to say, the takeover by the next, probably also untrained, Provo Sergeant was welcomed with open arms.

Guard duty with an unfamiliar automatic gun was not my only experience with a weapon different from the usual issue. This took place when to my astonishment, I was designated to be a sniper. Rather than with the standard issue semi-automatic M1 rifle, I was assigned a bolt-action single-shot rifle. This came after we had our first time at the firing range; and by accident, I did fairly well. We lined up in a rather ragged row, with soldiers with some experience standing by, giving firing-line advice to us mostly inexperienced recruits. Then, targets were raised and lowered, along with a kind of chant from Black soldiers who, from protected target pits, were in charge of checking the marksmanship of White soldiers. While Black Americans had been denied service as combat soldiers, they were allowed to function in some non-combat duties. Among the tasks allowed them at Breckenridge, raising and lowering the targets was as close to actually firing a weapon that they achieved while in the American Army in WWII.

From behind the protected target pits, the Black soldiers raised and waved a small flag while chanting, "The flag is up." Then, "The flag is down," as that small flag was raised and lowered from the pits. Our mentors then ordered us to "commence fire," after which we blazed away. I hit the target mostly by accident; but some others, also new to firing rifles, did not do so well. If you missed the target entirely, a rather strange red-patterned banner was hoisted, and your shame was revealed, with the chant from the target area for all to hear: "MAGGIE'S DRAWERS!"

In spite of never having fired a gun before, my accidental on-target score merited my appointment as the sniper for the battalion. Instead of the standard weapon, the M1 - an automatic rifle, but not too accurate a firearm - I was given a more accurate, and better made, single-shot World War I Springfield bolt-action rifle with a sniper scope. After every shot, the bolt had to be pulled back, ejecting the spent cartridge, and then slammed closed, placing a new bullet in place. The single-shot Springfield was slower than the automatic rifle, but much more accurate.

I did get in some target practice; but the fame as a sniper did not ensue, as the greater need, evidently, was for the very expendable rifle squad leader. The aged, infirm, and generally unqualified recruits in the 98th Division outnumbered those few of us still mobile and with some basic skills; and so I was plucked from future fame as a sharp-shooting sniper. With less than a month in the Army, I was given a job as an infantry squad leader, with the rank of Corporal, and eventually, Sergeant. The squad, consisting of twelve GI's (*government issue*, the slang term for "common soldier") was the basic element of the infantry. Besides the squad leader, there were the assistant (acting corporal), the BAR (Browning Automatic Rifle, a sort of portable machine gun) man, the assistant BAR man, who carried a considerable number of ammunition clips, and eight riflemen, who also carried grenades and often a device for launching them.

At Camp Breckenridge, with the rifle squads in place, we began to attempt to learn what it was like to be a soldier. Monotonous close-order drills, arduous long-distance marches, sleeping in flimsy tents - often on wet ground - and living on tasteless rations, I found out were some of the fun-type activities that you get into when you are not being safe for your mother.

# My Home for Three-And-A-Half Years
## The 98th Infantry Division

WHEN I JOINED THE Army, I was assigned to the 98th Infantry Division, an organization that previously had existed only on paper. It was activated fairly late in the war because up to then, additional manpower was not thought to be necessary. However, the dire casualties so far in the war against Japan showed the need for additional troops, particularly with the projected invasion of the Japanese homeland in the offing.

Because the 98th was technically located in upstate New York, it was called the Iroquois Division, in honor of the Indian Nation that had been so important in that area. To continue the identification with the Iroquois, the shoulder patch was a bright red Indian head, on an iridescent blue background.

The Division had never been activated, and so just about everyone connected with the 98th was new to the military world, or was a retread from long-ago conflicts. Our regimental commander, Major William Ransom, introduced himself by saying he had seen action in World War I, when soldiers were called "dough boys," and said we should think of him as "Dough Boy Bill." He went on to tell us in an assembly that "there are three kinds of infantry: mechanized infantry, motorized infantry, and the kind that just walks." He waited a minute for that to sink in, and then said, "The kind that just walks – that's you." He was not joking, as we were basically foot soldiers for three-and-a-half years. The word that came down was that there was a shortage of trucks, as the European campaign would need a lot of them, and we were not seen to be high on the requisition list for anything.

As was the case with most of the officers, Dough Boy Bill had not had any recent experience with the military. Some, known as "90-day-wonders," had graduated from Officers Candidate School; but others had not even had that experience, having gone directly to the 98th from college ROTC, where they at least had learned how to march, but little else. The non-commissioned officers, sergeants mostly, had been in the army for only a few months before being sent on to the 98th, to pass on what they had learned in that brief time. It was not an overwhelmingly experienced cadre.

And, my fellow trainees were not exactly an elite force. They were mostly draftees who had been exempt from, or rejected by, the draft because of their age or physical disabilities, and who had now been scraped from the bottom of the draft barrel and inducted into the army; plus a few of us eighteen-or-so-year-old recruits. The reason for this last-ditch participation in the military, of so many exempted or rejected, was the projected need for many more troops in the Pacific war.

Known as the "Old Men," a fairly large group of draftees, 36 to 45 years of age, were, for the most part, not in very good physical shape. At this time, the life expectancy for men was age 50, and most of the Old Men were close to reaching their physical life expectancy. They were drafted, however, because they were still single. Due to the lack of decent paying jobs during the Great Depression, they had never been able to afford marriage, and were thus available for the draft.

There were others with limitations: some were illiterates and others were partial cripples. Some of the Old Men were veterans of WWI. In addition to my friend, Barney Norton, there were several other veterans of the war in France, and one standout from the First World War, Norris Cotton, a dirt farmer from Maine. For physical reasons, he had been rejected by the draft in WWI. This time, the draft board had scooped him up, and it was a mistake; they had been right the first time - old man Cotton was not cut out to be a soldier. After many years behind a horse-drawn plow, with one foot up and the other down in the furrow, he could not keep in step. Whenever we were practicing close-order drill, shouts could be heard, "Get that man in step," whereupon we would all switch steps to be in synch with old man Cotton. It was no use; in a few steps he was out of step again. Eventually, Cotton was given any duty that would keep him away from the parade grounds.

Under the circumstances of the general disabilities of my fellow soldiers, my rapid promotion to Provo Sergeant of the guard and rifle squad leader with the rank of corporal was not a big deal. There were just not too many able-bodied men who could read and write, who were not crippled in some way, and who were otherwise available for anything more than being just a private. The combination of untrained officers and non-coms, with draftees who were in good part physically inept, caused some consternation in higher quarters, and eventually mandated specific training to overcome the deficiencies.

## CAMP BRECKENRIDGE

RICH FARMLAND ALONG THE Ohio River had been taken over by the War Department to create Camp Breckenridge. Like most Southern Army posts, it was named for a Confederate General: in this case, John C. Breckenridge, the Confederate Secretary of War, and a hero to the local Dixiecrat politicians. 36,000 acres along the Ohio River had been taken over to create this large training camp, which was still under construction when we arrived. Underfoot was a sea of thick mud, and our flimsily constructed barracks afforded scant protection from the weather. It was in the cold part of winter when we arrived, and a coal stove, fueled by noxious-smelling soft coal, supplied whatever little heat there was; but it was inconsistent, providing either too much or not enough heat.

The outstanding feature of Breckenridge was mud, which was probably an excellent ingredient for growing crops, back when Breckenridge was fertile farmland. It made military training difficult, as it was very sticky, clinging to our equipment and ourselves. A pail of water and a scrub brush were available at the entrance of every barracks, as we had to be scrubbed of the clingy mud after every excursion. It was a constant problem. Every evening, our equipment - weapon, packs and clothing - required scrubbing. As the camp was in Kentucky, satirical comments were often made comparing Kentucky's famous - but nonexistent here at Breckenridge - blue grass with the ever-present mud. But as time went on, and scrubbing continued to always be necessary, the joke became hollow.

About 100 cots were jammed into the barracks and scrunched closely together. The cot next to me was occupied by one of the 'old men,' Barney Norton. In spite of our age difference, Barney and I quickly became friends. He liked to talk about his World War, the first one, and I liked to listen to his stories about a war I knew little of. His genial nature made it hard not be friends with Barney; and as our cots were side by side, it was important to me as a new recruit that we remained friends.

Barney's slight build made his uniform seem too large for him; but around Breckenridge, uniforms that fit were not much in evidence. Slightly bald, with a prominent nose, and a mischievous twinkle in his faded blue eyes that competed with his often dour expression, it was difficult to guess which emotion he was expressing, or whether he was leading you along with a story that might be somewhat exaggerated.

"Old Soldiers Never Die" was a favorite ditty; and he used to ramble on about being one, often ending a conversation about his old army career by singing about the old soldiers in a high-pitched, slightly cracked voice - particularly when he'd had a bit of what he called 'the sauce.' Barney's involvement with the sauce, which he usually kept in his footlocker, did not seem to affect him in a negative way. Although, when he was in the sauce, he was apt to repeat his comments about his army career at inconvenient times, like when we were supposed to be at 'attention.'

In spite of his repetitions, and the odd times when he talked, I was pleased when Barney enlightened me about life in the military. His repeated stories were worth following, as they varied in each telling, with new details arriving with each interesting episode. Even though these stories often happened after taps and 'lights out,' I did not mind, as they were insights into how wars actually transpired, at least as Barney saw it.

His stories, often expressed in colorful language, included anecdotes about serving under Black Jack Pershing, in the pursuit of Poncho Villa in New Mexico, Texas and Mexico. General Pershing had received his nickname because many of the soldiers under his command, there in the Southwest, were Black. However, as soon as Woodrow Wilson, an unreconstructed Southerner, became President, he segregated the American Army, as well as the Post Office and all Federal buildings - starting with the White House.

After serving under General Pershing in the Southwest, Barney again served under Pershing in France, in the now all-White American Army. One of his stories about military action concerned his shock on seeing the bodies of dead American Marines in Belleau Woods. Here, the Army reinforced the Marines, who had suffered heavy casualties as they charged the strongly defended German machine gun positions, armed only with rifles.

When an American, Hiram Bingham, invented the machine gun, he offered it to the American War Department, which rejected his invention because machine guns had not been used in the Civil War, and therefore must not be important. When offered to Europe, all countries there eagerly accepted this new weapon. The English were so impressed that they not only made Bingham a citizen of Great Britain, they knighted him; and the German High Command built their war power around Bingham's invention. Early in WW1, the machine gun had become a most feared weapon, causing so many casualties that machine gunners, when captured, were quite often shot.

Barney had been involved in other battles in France, including heavy fighting in the Aragon forest; had been wounded, gassed and shell-shocked; and after the war, received treatment at Veterans hospitals. But, the honor attached to Barney's prior service did not deter his draft board from calling him up again in 1943, at age 44.

This was a different army than the one he had experienced so many years before, and it took some getting used to. Life in this new situation was somewhat of a shock to him - and to his no-longer-youthful body. But at least he had the experience of his former service to fall back on - and his supply of the sauce, if all else failed.

There were a few other WWI veterans who had been recalled, including Sol Bernstein from Boston, who got up to the front lines in France; but there was hit by shrapnel and invalided home. There were some other WWI vets among the 'old men,' but none of them had stories like those that Barney came up with.

## Infantry Training, Camp Breckenridge

For the few of us young soldiers, the infantry training was tough, particularly with the heat of the Southern sun bearing down; for the 'old men,' it must have been a nightmare. Calisthenics were performed in a large circle, on the double, with jogging interspersed with duck-walks and bear-walks; and we endured these oppressive exercises over long periods of time. The very active bayonet practice was squeezed in between hand grenade practice and judo. Grenade practice involved using dummy grenades that had a weak explosive force, and later, with grenades that could be lethal - and sometimes were, even in practice. Weaponless hand-to-hand combat for attack, and self-defense with judo were crammed in wherever possible. To get us used to live ammunition being fired close to us, we had to crawl on hands and knees with live machine guns firing over our heads as we crawled along. And in between everything else, we marched.

We had been warned that we were not the motorized, or mechanized, infantry, but that we were the infantry that just walks - which proved to be an accurate prediction, as we did walk, or rather marched, a lot. Sometimes we marched with light packs; but other times it was with a full field pack, which included a blanket, poncho, mess kit and field rations for several meals, a trenching shovel, medicine kit, gas mask, shelter half, and a tent pole and pegs. And, of course our weapons: the rather heavy M1 semi- automatic rifle, a cartridge belt loaded with live ammunition, wire cutters, and a sheathed bayonet. Additionally, we carried quite heavy, full canteens of water that were supposed to last all day, as part of the water discipline we were inflicted with, in spite of the fact that we were marching in the fierce Southern heat. In addition to our bearing rifles, every squad included a BAR (Browning Automatic Rifle), a quite heavy combination rifle and machine gun, which required extra ammo, to be carried by the BAR man and his assistant.

As the summer wore on, bivouacs of various lengths became common. These usually included several days of strenuous activity. To arrive at the bivouac assembly area, we often marched a number of miles; and as they were usually part of an exercise that lasted from several days to a week, that meant sleeping on the ground, for whatever time the exercise lasted. Because of this, we were encumbered with full field packs, and in addition to the packs, helmets and gas masks. We always carried our requisite weapons and all of the other appendages prescribed by the military. Once we arrived at the place where the military exercise was to take place and the fox holes dug, the next step was getting ready for whatever sleep could be obtained on the hard and often damp ground. The first step for that began with assembling the two-man basic army pup tent.

You shared your shelter half with another soldier - in my case with Luke Spaulding from New Hampshire. A guy about my age, he and I would work together in setting up the small tent. He had the other half of the pup tent combination, which was a matching shelter half and also a folding tent pole and pegs. The two shelter halves would be snapped together, tent poles would be extended and set up at either end of the combined shelter halves; the pegs were pounded in, and the tent ditched, to direct any rain away from our tent. One poncho was placed on the ground to keep out dampness, and a blanket would go on top of the poncho. The two soldiers would lie on that blanket, covered with the other blanket, and the second poncho would be the top cover; and the pup tent arrangement would be complete - our shelter from the night air. Not exactly happy camping, but comfort is not something to be expected in the military world.

You would, of course, sleep fully clothed, except for your boots, which, padded with whatever came in handy, would serve as a pillow. We usually rose at dawn, awakened by the cold – as dawn is usually the coldest part of the day - to look forward to another day of fun and games in the infantry.

One of our first marches - while a relatively short one, only fifteen miles – was difficult, as it was a first-time march in the summer heat. Difficult even for those who were younger, and very tough for the 'old men,' with one exception: Private Lindsey. Toward the end of this first long march, the rest of us were bushed; and, although he was 49 (old for those days, when life expectancy was age 50), Lindsey ambled along even in the late stages of the march as if he did not have a care in the world. His secret source of marching strength? In civilian life he was a trapper, and walked a long trap line every day; so fifteen miles was a stroll to him. Lindsey actually enjoyed army life: three squares a day, easy duty and 3.2 beer at the PX. What more would a man want? Lindsey was planning to stay in the army after the war was over. Army life was, for him, a piece of cake.

One of the toughest exercises of that summer's difficult training was a forty-five -mile march. This was a twenty-four hour, all-day-and-through-the-night excursion, encumbered with packs, gas masks, rifles and steel helmets. There was one three-hour break, and the rest of the time we marched. Nobody sang or counted cadence - we just plodded on, although at times you could find yourself falling asleep while marching. On our return, completely exhausted, we turned in, leaving a mess on the barracks floor. Shortly after, an inspection took place, and we were gigged because of the mess we had left; which meant no pass for that weekend; but most of us were content to stay quietly in the barracks. The 24-hour march took some time to recover from, and we needed that time to recoup our basic energy.

One early fall bivouac ended up in a yet-to-be harvested pumpkin and corn field, with the corn stalks tied up in stacks all over the field, and the pumpkins spread out between the stacked corn stalks. The entire scene made a very picturesque expression of autumn, down home on the farm. Spaulding and I decided to use the stacked corn stalks as a makeshift kind of tent. The trouble with this concept was that the harvested stacks, while loose at the bottom, were tied together at the top, using a center corn stalk, still rooted in the very hard soil, as the anchor for the assembled stacks. Unfortunately, the rooted stalk in the center of the corn stack kept us from sharing blankets. That night I was cold, as was Spaulding, and we never tried to create a corn stack sleeping arrangement again.

On this bivouac, we were in a defensive exercise, and we dug foxholes to protect against an attacking tank force. When the attack happened (and I don't remember who won), we first retreated; and then, when we reoccupied the former corn fields, it was evident that the tank crews had had some fun in this battle, as it was obvious that a good many of the pumpkins, courtesy of the tank crews, had become squash.

## An Infamous Bivouac

PROBABLY BECAUSE WE WOULD shortly be on our way to Tennessee maneuvers, where we would be mostly living in the field, bivouac became a frequent exercise. As these were usually two - or three - day events, we went out with heavy full field packs, and other equipment necessary for camping out - the military way. Included in the pack were a blanket, a raincoat, half of a pup tent, a tent pole and several pegs, a mess kit, and emergency rations. In addition, we carried our weapons - rifles or Browning Automatic Rifles - a gas mask slung over our shoulders; and at our waists, an ammunition belt loaded with bullets, a medical kit, a wire cutter, and a bayonet. Loaded down and hot, marching in the summer sun, we were usually relieved when we reached the designated bivouac.

After marching some distance from the barracks, we would be positioned in a spot that had been selected as our bivouac area. It was usually rather rugged terrain, where, as part of the military exercise, we would be assigned a place that we were to defend, or use as the jumping-off place for an attack. Digging foxholes was usually the second order after arriving, the first being the digging of latrines. Then pitching tents - or rather, shelter halves shared with a buddy, and fastened together to make a completed pup tent - and finally, eating some canned rations completed our day. After the march with heavy packs, and then digging in, we were usually ready to retire to the pup tents we had assembled, for whatever sleep could be obtained on the always hard ground.

On one bivouac, we camped in a different kind of area. It was fairly level ground that was obviously a recently used cow pasture. While it was no longer used for pasturing, as it was now part of the Camp Breckenridge acreage, it evidently had been heavily used by cattle, judging from the amount of partly composted cow manure littering the field. In addition, the swarms of flies that flocked around the manure attested to its former usage. As this was to be a battalion bivouac, a mobile kitchen had been set up in the pasture; and in spite of the swarms of flies that we had to keep brushing away, the hot food was enjoyed. And we all agreed that it was much better than the canned rations usually offered in bivouacs.

However, the elation was not lasting, as shortly after eating, many of us became desperately ill. The reaction for some of us to this calamity began with a rush to the recently dug latrine. But for others, it was more serious, and I was one of those who dropped to the ground, unconscious. Those of us who collapsed from acute dysentery were trucked off to a newly situated field hospital. The affliction we suffered from was just like the dysentery soldiers suffered from, and often died from, in the Civil War. I too would have died, as I had become completely dehydrated. However, once in the hospital, I was treated by medical personnel who relieved the dehydration that was at the point of doing me in. They did this through intravenous feeding, which saved me. But, during my drawn-out recovery, I could see why so many Civil War soldiers, in the absence of modern medicine, died from illnesses contracted in the field.

After a few days in the field hospital and a week in the base hospital, I was sent back to barracks to be on light duty. Before this episode I had been thin, but because of this illness, I lost 20 pounds, a weight loss I could ill afford. For quite a while, my insides remained shaky, and I did not seem to be getting over my malady. But, along the slow road to recovery, a homeopathic medicine was recommended by one of the medics - Tincture of Belladonna, that I was able to obtain from a local drug store, which helped my insides quiet down. I was later told that Belladonna (beautiful woman), which did soothe my insides, was made from a poison, Deadly Nightshade. In spite of its sinister ingredients, Belladonna was not fatal, but instead aided in my recovery. Eventually I was cured and restored to active duty; and shortly afterward, we were on our way to the Tennessee maneuvers - where the 98th Division would be evaluated for its fitness to be committed to an actual battlefield.

## I Become the Poison Gas Non-com for the Third Battalion

IN THE FEW SHORT years between the two World Wars, the lethal effect of poison gas in World War I had not been forgotten.

# I Become the Poison Gas Non-Com

Even before the U.S. was blindsided by the attack on Pearl Harbor, preparation for protection against poison gas had begun.

Gas warfare depended upon surprise to obtain its maximum effective use. Not long after the initial surprise attack at Pearl Harbor, all American soldiers were issued gas masks, with instructions that they must always be kept with the individual soldier. In addition, for every infantry battalion, a gas non-com was appointed, with instructions to learn as much as possible about the lethal properties of the various gases and be prepared to share this information with the troops.

Because of this specification, I was transferred temporarily from my position as a rifle squad leader to that of platoon sergeant and was appointed poison gas non-com for the Third Battalion. This was not the first time I was placed into special duty, nor was it the last.

In my new capacity as a gas non-com, I learned technical details about the properties of poison gas, whether released in the air, like phosgene, which smelled like moldy hay, or spread on the ground like mustard gas, which caused crippling injuries through physical contact. As gas non-com, I learned not only the technical properties of poison gas, but something of the strange, almost science-fiction utilization of this lethal weapon in the late Great War.

In addition to the technical details of the death-dealing chemicals, I was given information about the sinister background of the creation and development of poison gas as a weapon in World War I. I had already heard about poison gas use in World I from my friend and fellow soldier in WWII, Barney Norton, who had been subjected to a gas attack in "The War to End All Wars," World War I.

At that time, the back-and-forth, unwinnable trench warfare had caused a stalemate that was bleeding the armies on both sides, with no end in sight - until a group of German scientists from the I. G. Farben conglomerate came up with a way to end the stalemate: poison gas.

The scientists suggested this weapon as a way to break open the enemy's defenses and allow victory for Germany. This lethal weapon had never been utilized, as artillery shells containing poison gas were forbidden by the rules of war. To circumvent that restriction, instead of artillery shells, the scientists suggested that canisters containing gas could be opened near enemy lines, allowing the gas to drift over allied trenches.

When this unconventional weapon was first proposed to the German High Command, they rejected it. These conservative, aristocratic officers' ethical views, in those pre-Nazi days, were not much different from those of their medieval ancestors; and they opposed this bizarre proposal as being foreign to their values.

However, continued losses, and a serious shortage of food in Germany, eventually made this way of ending the war something to consider. So, when the scientists returned with the suggestion for a test case - a poison gas attack on a limited basis, in order to judge how effective this new weapon would be - the High Command reluctantly agreed.

A section of the Allied front defended by French Colonial troops and Canadians was selected and a gas attack was launched and carried out under the scrutiny of the High Command. The attack succeeded beyond expectations, with almost the entire Allied front wiped out. If the German military establishment had been prepared to take advantage of the effectiveness of this new weapon at this point, Germany could have won the war.

The Allied troops, badly crippled by poison gas, had collapsed, leaving an almost undefended front. However, the now empty trenches were reeking with poison gas and were impossible to occupy without protection from the fumes. The German High Command, not entirely behind this new weapon, had not prepared for a breakthrough. Gas masks or other protection were not available for the German soldiers to use in crossing over to the now abandoned, but impossible to occupy, Allied trenches.

In a memorable painting by famous American artist John Singer Sargent, a long line of Canadian soldiers, blinded in a gas attack, are being led away from the front. Their eyes covered by bandages, each soldier holds one hand on the shoulder of the soldier in front of him. The lead soldier in the long line was necessarily a soldier who could still see. Other than the lead soldier, it was a classic case of "the blind leading the blind."

Improvised gas protection was soon created by both sides; and in a relatively short time, professional gas masks and other protective measures were in use. The short-term advantage had vanished, and poison gas became just another weapon in the squalid world of trench warfare.

Eventually, the ability of the Allies to also use poison gas as a weapon - and their new-found protection from the effects of enemy gas attacks and their ability to retaliate in kind - made the use of poison gas almost entirely obsolete. However, considerable damage had occurred, and for some years following WW1, Veterans hospitals continued to treat a number of victims of poison gas warfare.

After World War I, edicts against the use of poison gas were declared. But when WWII broke out, the Nazis, who disdained all treaties, might well have considered using poison gas. And so, to protect against this possibility, gas masks became standard equipment for the American military.

By the time we had finished Tennessee maneuvers and were on our way to our next destination, the threat of gas warfare had greatly diminished, particularly since the Allies had taken control of the air over Germany. As there was now little possibility of its use by the Nazis, there was now no longer any point for further protection against poison gas. From that point in time, gas masks were no longer a burden weighing down soldiers, and were collected to be available just in case they were ever needed again.

Gas non-coms were no longer needed either; but I was thankful that I had been able to learn the history of poison gas, which my friend and fellow soldier Barney Norton first told me about, as he had undergone gas attacks in France in WWI. The use of poison gas in warfare is still forbidden, at least, officially, for now.

## THE 98TH AND ROMMEL'S AFRIKA KORPS

WHEN I WAS IN the Army at Ft. Breckenridge, we were near a prisoner-of-war camp for Rommel's Afrika Korps. Hitler had not allowed them to be evacuated to Europe during the last days of the North African campaign, saying they must fight on and never surrender. They finally did surrender, however, after being overwhelmed by American and British troops, and many were brought to prisoner-of-war camps in America. Marching past their prisoner-of-war enclosures on our various military exercises, it was plain to see that these were not common soldiers, but obviously were the cream of the German youth. In perfect physical health and burnt almost black by the desert sun, they looked askance at us laggard city types, mostly in poor shape, as we limped and staggered under heavy packs past their enclosure.

Allowed to work outside the prison camp, the German prisoners marched in perfect precision, singing as they marched, like a trained operatic chorus. It was pretty scary, as we were not very good at anything military, and they were so good. It was evident that, at that time, they were still super Nazis, showing us what we were going to be up against.

In the end, with the Nazis in retreat everywhere, General Rommel had escaped North Africa with a sizable contingent of Afrika Korps soldiers, in spite of Hitler's injunction to stay on and fight to the death. As the Nazis were no longer winning the war, Rommel, like many other German aristocrats, now became part of a plot to affect a truce with the Allies, which included assassinating Hitler. The plot failed, and Rommel, implicated, was sheltered from the S.S, by his Afrika Korps troops. The S.S then told him that if he did not commit suicide, they would kill his son. Rommel complied.

Things had changed here at the prison camp, now that the American and Russian armies were closing in on Hitler. The United States government had gone looking through the prison camps for prisoners who could be part of the "new Germany." As that would probably be happening soon, we were allowed to talk with some of these ex-Nazis about life under Hitler, and what they thought the future might be when the Nazis were finally defeated.

However, for those of us facing the powerful Japanese armies, it was a different story. The Japanese still had a large and powerful military force, and a civilian population that was ready and willing to die for the Japanese Empire. Invading the heavily fortified Japanese homeland would be a bloodbath for both sides.

## Yoo Hoo Ben Lear

IN WORLD WAR II, American armies overseas were designated by numbers, starting with the noted First Army. On the other hand, while in training, all American soldiers automatically became part of the 2nd Army, then under the command of Lieutenant General Benjamin Lear. That of course included my outfit, the newly developed 98th Division, which endured, like all 2nd Army troops, strange quirks in the basic training requirements that were created by General Lear. A stern 'Old Army' officer, Lear rigorously enforced discipline and had never been known to smile.

A story that went around implied that General Lear was sent back from Europe to be in charge of the training of the fledgling troops of the 2nd Army, in spite of the fact that he was supposedly involved in the failure of American troops to overcome the Nazi forces at Kasserine Pass. Once back in the States, his strange concepts of morale building and training tactics for the 2nd Army did not make early days in the service easy to take for anyone, including those of us in the 98th division.

General Lear was handicapped not only by his strange concepts, but by his nickname. In 1941, during the peacetime draft, while playing golf in civilian clothes at his country club, not far from some young women golfers in shorts, he became angered when some soldiers in a passing convoy "Yoo Hooed," at the young women. Outraged at this crude action, General Lear, noted for his grim demeanor, jumped into his jeep, roared up to the head of the convoy, and ordered the entire 35th Division to get out of their trucks to march on foot the remaining 15 miles to their destination.

Most of the soldiers in the convey were recently drafted, inexperienced troops, and they suffered on the march, with some collapsing in the 98-degree heat; but that was nothing compared to the heat General Lear suffered from his empty-headed vacuity. The word of this nonsense soon spread. So, in addition to much immediate negative publicity, for the rest of army life he was known as Yoo Hoo Ben Lear.

General Lear had interesting concepts about training and in the building of morale, some of which he inflicted on the newly created

98th Division. As a presumed morale builder, he went overboard in his enthusiasm over Irving Berlin's hit song "This is the Army, Mr. Jones." The chorus detailed Mr. Jones' deprivations since he was summoned by the draft, as the poor fellow now had," no private rooms or telephones." The song detailed the privations of Jones and others, whose nice, easy lives were changed by the draft. This lightweight nonsense was a favorite song of General Lear, and he decided that it would be a positive morale factor if every member of the enormous 2nd Army learned to sing it. This was accomplished by requiring every soldier to sing, or at least mumble, the entire song before being allowed a pass to town.

The 98th Division was composed of diverse ethnic backgrounds, with myriad cultures and resulting accents. It was interesting to hear the many linguistic versions of General Lear's favorite song, as the troops complied with the assignment required for a pass. Among the ethnic diversions was Alexis Kostopolous, one of the older recruits and a recent arrival in America, who spoke a Greek- English dialect exactly like a popular radio comedian who told jokes with a broad Greek accent. Kostopolous was a lost cause as a soldier, and he was usually relegated to trash detail along with Bootwell, an older farmer from Vermont who also couldn't march or shoot straight. Because of their lack of ability to throw dummy hand grenades in training any distance at all, both were strictly forbidden to have anything to do with live hand grenade practice.

When it was Kostopolous' turn for an application for a pass, this occurrence drew an audience to the orderly room to listen to him mangling, in a wild Greek accent, Berlin's Mr. Jones, etc. He was not the only foreign-accented applicant to mangle the required song. There were many, including Abdul Bashir from India, and an Indian from America, Glen Tsosie, a Navaho, who in addition to being completely illiterate, spoke no English before being drafted. Not every singer was tone deaf. but many were, and everyone had to sing to get a pass. A variety of Southern accents, contrasted with those of Brooklyn and the lower East Side of New York, created a wide variety of regional accents in their renditions of the adventures of Mr. Jones, Mr. Green and Mr. Brown, who were now Privates Jones, Green and Brown, as if anyone cared.

We never did find out the full story of General Lear's connection with North Africa, but it was supposedly from the British there that he became aware of the North African water discipline policy. This rigorous system, according to the information reaching General Lear, allowed the British soldiers a single canteen of water, about one pint, per day for all uses -- washing, drinking, and all other needs.

If the British could endure this in North Africa, General Lear reasoned, American troops in training could do the same. And so a strict water discipline regulation that permitted American soldiers in training only one pint of water per day, was stipulated for the Second Army in the hot Southern summer.

It was possible to get by while in the barracks, but tough to endure and hard to enforce in the field while training for combat, a time when resupplies were not allowed. While water discipline was pretty much enforced, there were ways around it, and exceptions took place. On any march or maneuver, the Army allowed a ten-minute break every hour. When the break occurred in the field, water carriers, usually fast movers, were appointed to take clusters of canteens to available water supplies, usually from the wells of abandoned farms that had been taken over by the government to be part of the grounds of Camp Breckenridge.

Sprawling over 36,000 acres, Camp Breckenridge was created from prime farmland that had housed over 2,000 residents in 552 individual farms. While the houses where the farm families lived were torn down when the government took over the land, shade trees marked the place where farms had been, and wells still remained. The trees were our beacons that pin-pointed the location of wells. Even with the abundance of abandoned farm sites, ways to replenish our rationed water supplies were not always available; and the one canteen of water a day in the hot Southern climate caused much suffering. However, as we were forced to comply with the water discipline rationing, over time it did toughen us, and by the end of summer we were able to suffer through our ration of the one canteen of water per day.

On our last bivouac before shipping off to the Tennessee maneuvers, all troops were forced to participate in a final Breckenridge exercise - even the quartermaster clerk - a healthy young man who had never been out of his company headquarters office or on a maneuver that featured water discipline. At one point in the exercise his natural thirst overcame him, and he practically passed out, flopping face down in a shallow pool of muddy water that troops had been marching through. He was quickly rescued by the medics, but his collapse showed that practicing privation could make a difference. Even though we eventually learned that our privation had been unnecessary.

As the summer wound down, and we were getting ready for the big challenge, Tennessee maneuvers, the emphasis on water discipline ended. Shortly, the full story of the North African water discipline leaked out - it had been a fiasco. First, it became known that the British had tried water discipline in the winter, when North Africa could actually be quite cool - as opposed to the brutal heat that our water discipline program had been implemented in. The second part of the story was that even in the cool winter weather of North Africa, the British had discovered that one pint of water a day was nowhere near enough, under any circumstance. Shortly after they instituted water discipline, the Brits found out that it was an unfeasible concept and abandoned it. However, General Lear's North African connections had not apprised him of this fact until our brutal summer was almost over.

It was not that easy to say goodbye to Irving Berlin's Jones, Brown and Green and their sponsor, General Lear. As the 98th Division was about to leave Breckenridge, and the 2nd Army, for Tennessee maneuvers, Irving Berlin's trio still lingered on in our Ben Lear culture. It was the main feature of our goodbye parade to General Lear and basic training. After having had to learn the words to be approved for a pass to town, we now had to sing it one last time. A stand had been set up on the parade grounds for General Lear to review the 98th Division as it marched past, with every soldier singing, as we marched, all four verses of "This is the Army, Mister Jones." The number of times General Lear would hear that song repeated over and over might have been an infinite pleasure for him, but for others on the reviewing stand, as wells the marching soldiers, the constant reiteration of the introduction of Misters Jones, Brown and Green to army life was intolerable. We were certainly sick and tired of the banal thoughts expressed in the lyrics and hoped it would be the last time we had to listen to it.

When the time came for the final parade, it took some time to get everything ready for our part in the parade and song fest. When it began, we were lining up on a steep hill that led down to the parade grounds, when a commotion at the top of the hill caught my attention. Because the hill led down to the parade grounds, it was crowded with soldiers and their officers, waiting for their time to march down the hill and become part of the parade. My outfit, the third battalion was the last to march and we were ready to go down to whatever nonsense was waiting for us at the parade ground, when the commotion, caused by a loud rumbling noise drew my attention, along with that of everyone else. Surprisingly the hubbub was caused by the two non-marchers: Bootwell and Kostopolous.

One of the original concepts of the 'Goodbye Parade' had been to keep the non-marchers among us - a considerable number -- away from the parade ground. This included Bootwell and Kostopolous, whose trash pick-up routine would be allowed to continue, but well away from the parade grounds. Their vehicle was a two-wheeled cart, with handles extending from the front, by which the driver – in this case, Bootwell – pulled and pushed to propel the cart to its next destination. Kostopolous' job, up in the cart, was to pick up the trash containers handed to him by Bootwell and dump the trash into a container in the cart. The trash container when filled was taken to a central trash area. The clean-up groups usually were assigned a specific area, but now had been given a new route. Warned to keep away from the parade, the trash boys had become confused, and by mistake, turned on to the hill that led to the parade grounds. Once on the hill, because of the steep grade, there was no turning back as gravity took over. Bootwell had lost control, and the rumbling sound heard earlier turned out to be from the cart, now a runaway vehicle which skittered noisily by us as we waited there on the hill.

Bootwell, whose job had been to hold on to the front handles of the cart and drag it along, was now hanging on for dear life. Half of the time his feet were kicking in the air, as the front of the wagon tipped up; and the rest of the time, his legs were hitting the ground with no effect as the cart tipped up and back as it rolled erratically down the hill.

As the very unstable runaway cart, with its crew of two, rumbled past the troops lined up in formation on the hill; this massive assemblage of the military did not phase Kostopolous. He was unperturbed by the commotion, and the rough and rapid ride that left Bootwell desperately hanging on. As the cart rumbled noisily past the massed troops and the host of assembled officers, Kostopolous, standing in the careening cart, holding on to a trash can with his left hand, but mindful of military decorum, gravely saluted the officers, as the cart lurched its way down the hill, leaving consternation in its wake. As the runaway cart passed my squad, Kostopolous recognized me, but, he did not salute, as I did not rate one, but gave me a discrete wave; it was obvious that he had enjoyed his five minutes of fame.

At the bottom of the hill, the cart with its occupants was quickly escorted away to whatever punishment this breach of decorum entitled them to. However, no time was lost, the parade had to go on, despite being thrown completely off balance by the garbage cart debacle. When it was our turn, we scrambled into proper ranks and files and began our march down the hill to parade and sing for that rather strange Lieutenant General in charge of the Second Army, whose nickname and strange military training methods persisted, I assume for the rest of his career. But we never knew what happened to either Bootwell or Kostopolous.

As far as the rest of us were concerned, with our rugged basic training, water discipline, Irving Berlin's songs, wayward trash carts, and Yoo Hoo Ben Lear, all behind us, we soon were on our way to maneuvers in the Tennessee foothills.

# Part Two:
# Tennessee Maneuvers

## Tennessee Maneuvers

AFTER ABOUT SIX MONTHS of intensive training, my outfit, the 98th Division, was sent south to undertake Tennessee Maneuvers. Here we engaged in mock warfare with two other infantry divisions and an armored division. The maneuvers were pretend battles to see if these outfits were ready to be transferred overseas. As there had been some retrenchment in numbers in our division due to physical problems, new recruits were added to our ranks. These were mostly young men who had been given a few weeks of training and had been spared the intensive toughening that we had been subjected to.

The mountains of Central Tennessee were rugged, and as it was late in the fall, the combination of rugged terrain and rather miserable weather made things even more difficult. It often rained, and nights were always clammy and cold. Although some of the physically maladjusted soldiers in our outfit had been discharged, there were many more who were barely making it. In fact, there were still a lot of what we considered "Old Men" soldiers, who because they were in their 40s, had a life expectancy of age 50. Life in the infantry is always difficult, but for them, it was pretty close to impossible. But they were not the only ones with problems. The 98th Division was new, with many bottom-of-the-barrel, unfit-for-combat soldiers in our ranks. And as we were lacking experienced leadership, things often went wrong.

We did not do well in the maneuvers, often losing whatever battle we were involved in; in which case our chow trucks were often captured, forcing us to miss meals. Fortunately, the Tennessee mountain people, who usually had family in the service, would take care of us. We could go to one of their simple country homes and request food, which they would then cook up for us for a small charge.

It was at this point that a Navajo, Glen Tsosie, was added to my squad. A new recruit, still very young, his military training was limited, partly due to the fact that he was completely illiterate and spoke no English (though he could count up to ten in Spanish). This did not seem to matter, and as he was in excellent physical shape, he quickly fit into Army life.

It was not that easy for other new recruits. Horton, a rather plump young recruit from Chicago, was not up to the rugged physical activity that we were accustomed to, and he broke down on one of our more punishing marches. Unused to the physical and mental stress, and completely exhausted, he dropped out of line, sobbing and miserable. This was just as our tough first sergeant, Dan Norton, happened by and quickly ended Horton's breakdown. Norton, in an old-Army tone of voice, told the new recruit that if he did not get back in line pronto, he would kick his ass all the way to Murphysboro (our destination, about 10 miles away). Horton immediately stopped crying, quickly got back in line, and never dropped out again.

Nashville, the important capital city in Tennessee, was important to us because if we got a pass and secured transportation to town, we could get to the "Y" for a shower. That was important, as there was no other way to keep clean. Hot meals in one of Nashville's restaurants was also much appreciated. But for the mountain boys in our outfit, getting to Nashville was as if they had died and gone to heaven, as Nashville was the point of origin for the famous hillbilly radio show, "Grand Ole Opry," a staple in their mountain cottages since they were little. Now they could go in person to view the Grand Ole Opry as it was being performed. It was hillbilly nirvana.

The Army provided trucks for the weekend excursions to Nashville, and Glen Tsosie, on our first excursion into town, rode in with the rest of us. But, never having been in a city, he got lost, and because of his lack of English, neither speaking it very well nor having an ability to read signs - missed the last trip back to our maneuver weekend headquarters. Tsosie was punished for missing the last truck by being restricted to our bivouac area, weekends; and while we were still able to get to our weekend shower at the Nashville 'Y', Tsosie kept himself clean by bathing in cold mountain streams that were near the weekend bivouac area - something most of us city types would not have been inclined to emulate.

As the 98th was a new and untested outfit, just recently created, the battle losses were a concern of the commanding officers, who cared about future promotions; and of considerable concern to the troops, as we had to contend with the rigors of rough mountain terrain - sleeping on the cold, and usually wet, ground in primitive two-man pup-tents, and huddling in them for shelter during the frequent rainstorms. And because we were usually on the losing side of the mock battles, we often suffered the loss of our chuck wagons. This left us without food, making the cold rainy days and nights even more miserable. Fortunately, the mountain people in the maneuver area usually took care of us. The local mountain people often had kinfolk in the service, and while they had very little, they helped us when they could.

## The Road to Possum Hollow

IN LATE SEPTEMBER IN Tennessee, by then cool and often wet, the 98th Division was fairly far along in our Tennessee maneuvers, but unfortunately not doing very well; we were losing most of our mock battles. In spite of our failures, the maneuvers continued. As part of one particular exercise, I was on a team that was assigned to create an outpost in one of the hills above Nashville. There were four of us, three of whom were southern GI's: Carlton Sizemore, from South Carolina; Ed Davis, from Georgia; J. D. Lumsden, from Virginia, and myself. My companions, country boys, and all good friends, were used to tramping around in mountain areas, and were otherwise used to backcountry living. That was important, as it wasn't easy to find our way up through rough country to the area where we were to establish the outpost. This was backcountry without many people living in it, and not much available in the way of education, or indoor plumbing, electric power, or any other 'modern improvements' for those who did live here. Nothing much had changed since the early days of this country - in fact a Civil War reenactment could have been staged there at that time without removing anything, except automobiles and trucks.

We headed out cross-country, as there wasn't a road, just a kind of path that followed a creek, requiring some fancy footwork because of the many rather slippery rocks along the way. Following this path would not have been easy, even for local people; but, as the path was pretty much overgrown, evidently few visitors came this way.

At one point we decided not to continue to follow the creek, as our mission was to establish a lookout for the "enemy"- soldiers from the division we were competing against in the maneuvers - reported to be lurking in the area. We scrambled up and down rocky mountain gullies and escarpments to find a place where we could somehow hang out, and there, watch and wait for the appearance of the enemy. When, and if, this took place, we would then scramble back down to make our report; or if nothing happened, to wait out the end of this week's maneuver before returning to our regiment.

Our patrol, climbing up into the hills, was away from any sign of human habitation. And, as our search was for a somewhat level area where we could set up our outpost, we eventually worked our way up a heavily wooded hill to its crest, an area that was rocky and cleared of trees. Looking out from the crest, we were surprised to see, in the distance, an isolated but obviously inhabited valley. In this valley, stretching out in a low area between the hills, was a working farm with cultivated fields. The farm buildings included a small rustic house, perched on a slightly raised spit of land, along with a weathered barn and some other outbuildings. Providing a scene of antique tranquility, it was like a postcard from the past; and it was, we agreed, a comfortable and secure place to wait for whatever was due to happen in this remote aspect of the maneuver.

The farm and its buildings were quiet, very quiet, almost as if no one was living there, which made it seem to be an almost perfect place to set up an outpost. The house looked deserted; but as we approached, a man, accompanied by his dog, came out to meet us. He was an older man, slightly built, with sparse grey hair, in faded overalls and a worn flannel shirt. With his face as weathered as his overalls, pierced by crinkly blue eyes, he was obviously the farmer who lived here in the valley. He seemed pleased to welcome us, as perhaps this isolated valley had few visitors.

It was late afternoon and the air was getting chilly, so we were certainly happy to discover this farm with a barn, as the barn was a possible answer to our problem of finding refuge from the cold night air. The late September nights in the Tennessee hills had become too cold for the one blanket we carried in our pack to keep us comfortable. So, after exchanging pleasantries, we asked if we could sleep in his barn. He cheerfully agreed to that, and then asked us if we would like to come in and sit by his fireplace for a while. We of course agreed, and on entering the house, we stepped back in time to another era. Obviously, not much had changed since the early settlements there in the valley, which, we learned from the farmer, was called Possum Hollow.

On the small, simply constructed house, weathered unpainted boards provided the outside walls, and an unfinished stone chimney protruding from the tin roof completed the rustic appearance. The plainly furnished interior consisted of two bedrooms, the living room with a fireplace, and a kitchen with a wood stove. There was no electricity or running water, except for that supplied by a pump in the sink. By the pump was a cup, which, as customary, would be filled with water every night to allow the pump to be primed in the morning. We were not surprised that there was no bathroom but were surprised to find that there was not even an outhouse; and that, in order to not waste anything, the garden by the kitchen door was the repository for human fertilizer.

As we stepped into the living room, the farmer introduced himself as R. J. Harden. We met his wife, Martha, a spritely lady with a winning smile, wearing an obviously handmade dress probably made from flour sacking. We were then introduced to Harden's elderly mother, who, wrapped in blankets, was sitting in a rather primitive wheelchair - obviously an invalid, but with eyes that sparkled as we were introduced. Although it quickly became apparent that she was not only no longer mobile, but she was also completely deaf, and, like her son and his wife, could not read or write. This made it difficult for her to understand what uniformed men, armed with rifles, were doing in her house - here in this isolated valley; and we could see that our presence concerned her. As a child, she possibly lived in this valley during the Civil War, but there was no way of explaining what was now going on; and as she could not read or write, communication with her was just about impossible.

The Hardens offered to share a meal with us, and we of course accepted. After the dishes were cleared away from the simple but delicious meal, the elderly lady's primitive wheelchair was moved over to the fireplace, and we all, including the dog, joined her there. We sat by the cheerful light of the flickering flames and did the thing country people often did: we told stories; and as Sizemore, who always carried a harmonica, brought it out, we sang songs. It was, I suppose, what went on in many of these unpretentious, backcountry households. And while it was a new experience for me, my fellow soldiers were right at home. Like most backcountry men, they chewed tobacco, as did our host, the farmer. His wife and mother were more gentile, dipping snuff, which seemed to be about the same as chewing tobacco; and the chewing and snuff-dipping went on while storytelling took place.

The tobacco use seemed to be part of the evening's activity. From time to time, the tobacco chewers and snuff users needed to spit; but it was part of an accepted routine that a story was not interrupted until it was finished. And then, and only then, the ending of the story was punctuated by everyone spitting into the fire, which sizzled when the multiple spits hit the flames. It was a scene from the distant past recreated exactly as it would have been in earlier times, and it was a comfortable and comforting scene.

When the fire began to burn down, we bid good night to our hosts; and at the barn, climbed up into the hay-filled loft to compose ourselves for sleep. Sleep, however, did not come easily, as families of field mice shared the hay with us and seemed to be active most of the night, running around, back and forth, doing whatever mice do in the night.

The next day, we joined the Hardens for a simple but delicious breakfast of ham and eggs. The eggs, from their own flock, and the ham, also local, were evidently an important part of their diet, along with locally grown corn and other vegetables they raised. According to my fellow soldiers, in that part of the South the hogs were not penned in, but ran loose, finding forage in the nearby woods. In the fall, when it came time to harvest the hogs, they were tracked down and shot. Most of the meat from the hogs was then smoked to augment the winter supply of hearty foodstuff.

We had just finished breakfast when, somewhat unexpectedly, up through a somewhat dry creek bed, the "enemy" arrived. To avoid capture, we stayed inside the house, keeping out of the sight of the soldiers outside. This must have been alarming to Harden's mother, but there was no way of explaining to her that this was just military training, and that, while there was a real war on - and that we would eventually be part of - this warlike activity was just a training exercise.

By afternoon, the enemy had left; and with our mission over, we bid goodbye to our newly acquired friends from this lonely, but heartwarming, valley. Leaving some money to at least contribute toward the food they supplied us, we then made our way back to our unit, leaving this enclave of the rural past, rejoining the present at the weekend assembly of modern military forces. However, my partaking of a way of life, much as if it had been in the remote past, was an experience I will always treasure.

## Country People

THE 98th DIVISION'S FORAY into the mock warfare of the Tennessee maneuvers was the War Department's way of testing the Division's capability for the real thing. However, it was apparent from the mock battles that took place that the 98th Division was not ready to be committed overseas, and a change of plans ensued. We were, it was said, to be sent to another camp for reorganization and further training; and that, as part of the reorganization, many of the over-age and unfit soldiers were to be released, and some officers replaced.

We foot soldiers involved in the maneuvers went through much discomfort in the marches and counter-marches in the inclement weather that took place. The missed meals due to our chow wagons being captured, along with the ignominy of defeat, were constant negatives. But in spite of the hardships and the military failures, we had done our best; and we now knew that we were capable of working together as a team.

From a non-military viewpoint, much had been learned about ourselves and about the different way of life of the people of the hills of the rural South. Because we were involved with them, and their lives, we were treated to an exposure to a way of life that was little changed from America's early days; and we got to know and share the lives of people who lived in an historic kind of existence that only existed in mountain areas such as those around Nashville. Folks here had little enough of material things but were always ready to do whatever they could to make things better for us.

An important element of the way we explored life in the mountains was through the many patrols into the backcountry. These patrols gave us the opportunity to observe, and be part of, the simplified way of life of these country people. On one such back-country excursion, I came across a scene similar to what could have been observed centuries ago, and yet was still happening here in the Tennessee hill country. On this patrol, we had passed by a field of unusual-looking leafy stalks; and I learned from my patrol associates, who were from the South, that this was a special kind of backcountry sugarcane, known as sorghum. The usual sugarcane needs a tropical climate to ripen, but sorghum thrives in a cooler climate, and so was found in this mountain area.

I had not heard of this product, but it was highly prized, I was told, when rendered into sorghum molasses, a product much like maple syrup. However, maple syrup comes from Sugar Maple trees that grow only in the North and is extensively produced. While maple syrup has many adherents, the much less available sorghum syrup is highly prized for its distinctive flavor.

Soon after discovering the growing cane, we came upon a small clearing that provided a demonstration of how sorghum cane had, for centuries, been converted into syrup. This conversion was effected using a primitive apparatus of a kind that had been in use since biblical times. In fact, the scene in the clearing conjured up visions of biblical illustrations, as it demonstrated a cooperative effort between mankind and an animal - in this case, a mule. While the scene was real, it had about it a kind of dreamlike quality. One end of a long pole, dark from many years of use, was attached to a grinding apparatus, and at the other end, the pole was hooked up to a rather elderly-looking mule. A small boy led the mule as it ambled in a prescribed circle, causing the grinder to function. At the grinding mechanism, an elderly man fed sorghum cane into the grinder as the mule walked along his circular path. A bucket, just below the grinder, was there to catch the resulting syrup as it was ground out of the cane.

This rather primitive-seeming production looked like a lot of work for not a lot of syrup; but when I later tried sorghum molasses on pancakes, I appreciated the effort to create this delicacy. And the visual enactment of an ancient way of life made a strong impression on me that remains vivid these many years later.

This was not my only patrol. There were others, like the one to remote Possum Hollow; and these glimpses into a way of life outmoded elsewhere were picture-stories revealing a different world. These country people seemed contented enough, even with their low-income lives and sparse educational opportunities. They had their music, much of which harked back to Elizabethan times, and was everywhere, particularly at their communal gatherings or church and other festivities. Recreation was simple, consisting mainly of horse racing. These country folk also had a quaint way of speaking, using picturesque words and phrases; their words, like their music, harked back to early days in America. Their food, mostly raised by themselves, although plain, was hearty and wholesome. In short, there was really nothing for them to complain about - except a lack of electricity.

However, life without electricity was probably in its last stages, as the electrical system created by the government's massive Tennessee Valley Authority would soon enough provide power, even to remote areas. But what we experienced was a rather incredible living history, that we were part of by just being there. And in spite of the discomforts we suffered in the Tennessee maneuvers, the experience was worth the effort.

As our Tennessee maneuvers ended, word got out that we would be sent to Camp Rucker in southern Alabama, for reorganization. We were soon off to the deep South, but a shortage of Army trucks necessitated a combination truck-and-march transportation episode that lasted three days. Alternating between marching and truck transportation, our three-day excursion to southern Alabama allowed us to see a lot of countryside along the way.

## Fort Rucker

AFTER HAVING FAILED MISERABLY in Tennessee maneuvers, the 98th Division was banished to the lowest depth of the deep South, Fort Rucker in southern Alabama. This was a time of reorganization, with many officers shunted off to other commands or desk jobs in some remote area, and new ones assigned. The cripples and other handicapped soldiers were reassigned to light duty or sent home, and the experiment concerning the drafting of "Old Men" for front-line infantry duty was quietly shelved. These elderly and infirm draftees were informed that working in a defense plant was more important than Army duties; and all they had to do was get a letter from a factory that made military equipment accepting them for employment, and they would be free to be honorably discharged. It did not take long before most of the 'old men' were on their way home. And, a few of the not so old ones were also free to go.

For one of the not so old men, George Koch, it was different, as military life actually worked to his advantage. Even though his health was somewhat shaky, the 98th Division actually helped George Koch turn his life around. In his first contact with the military, he had made a serious mistake, by accepting the summons from his draft board. He did not have to become Private George Koch. He could have evaded the draft, as he was in his late thirties, was married, with a wife and two young daughters, and he was not that well. With a family to support and with his health problems, he could have stayed out of the military. But for George Koch civilian life had been pretty crummy. His original good job had perished with the '29 Depression; and when he found work again, it was a boring job for lower pay than he had experienced before the Depression.

And for George, family life was discouraging; money was scarce; and with the economic problems that the family faced, his wife now had to work outside the house. Because of that, taking care of the children was difficult; particularly since a medical problem was evident. The girls, 10 and 12, were good kids; but his younger daughter had a serious physical problem, a severely crossed eye. To repair that problem required an operation that George Koch could not possibly afford, as at that time medical insurance did not exist, but without an operation, sight in that eye would eventually be destroyed. Family life, like everything else concerning George Koch, was not a happy life.

To get away from all of this George Koch had not contested his draft notice, but once in the service he now found that life in the infantry was more difficult than he had imagined, and at times more than he could handle. As the physical hardship promised to get worse, there seemed to be no way out of this morass, now and in the future.

That is, until his wife and daughters came for a visit to Fort Rucker. This opportunity for a change of fortune came at the end of basic training, a time when family visits were encouraged. As George told me, while walking with his visiting family on the way to the Post Exchange, they happened to meet Dr. Krause, the surgeon in charge of the medical needs for our battalion. George, while receiving treatment at the battalion hospital, had become aware of Dr. Krause's civilian practice as an eye surgeon; but he knew that the good doctor's special surgical ability had been restricted by his duties as a general surgeon. But, when the doctor was introduced to George's family, a new development in the world of George Koch emerged, as Doctor Krause immediately saw - not just a child with a serious eye problem – but a child with a problem that he could correct. And his optical surgical expertise and humanity immediately came to the fore. Dr. Krause offered to operate on the child's defective eye, with no charge for the operation. His offer was accepted. The operation took place, was successful, and George's seeming stupidity in allowing himself to be drafted turned out to be a very rewarding experience - particularly when it was realized that the attempt to utilize older men for the infantry was not working, and it was decided to release over-age infantry soldiers. When the 'old men' began to be discharged from military service, George, through connections with Dr. Krause, was one of the first to be honorably discharged. He went home to his now happy family; to a new and interesting job in one of the defense plants - and to a realization of how lucky he had been.

With maneuvers over and military training moderating, family visits were allowed; my sister, who lived in Indiana, visited, and an aunt from another part of Kentucky came by. Others had similar visitations and reunions, but not everyone had the experience of good times from family get-togethers. When I heard that my friend, J. D. Lumsden, was expecting a visit from his wife, I was going to offer congratulations, but on talking with him, it was apparent that he was not exactly overwhelmed with pleasure about this expected visit. As we were good friends, I waited until he was in the mood to talk about the expected arrival, and his problem with it.

J. D. Lumsden was worth knowing. He was a character, not just because his first and his middle names were initials, and his Southern dialect spilled over with quaint words and country sayings, all this, in addition to his often on tap, sense of humor. He was a born comedian who could make his recital of ordinary occurrences such as a trip to the Post Exchange into a hilarious happening. He could even make a recounting of a session on the usually irksome kitchen police recalled as an occasion for laughs

In spite of his age, 43, J. D. did not really qualify as one of the 'old men.' He was in good physical shape and was, in fact, a natural for his slot in the army. In every infantry company, three rifle platoons were assisted by a weapons platoon. This consisted of one or more light machine guns and a trench mortar. Working with this kind of equipment was where J. D. excelled. Because of field conditions, where heavy usage could cause problems, a company artificer was needed to keep the weapons platoon equipment functioning. J. D. excelled as the company artificer for I Company. This was a job description that suited him to a T, as he could use his experience to do what he liked to do, which was to make mechanical devices work.

Without much education, working in the only viable enterprise in his hometown, a rather ramshackle cotton mill, J.D. had become a kind of untutored hands-on mechanical engineer. There he coaxed semi-obsolete mill machinery to continue to function in spite of age and condition. He did this by assisting ailing and outmoded machinery continue to work by reviving old systems with bailing wire and tape, and inventing replacement parts; manufacturing what he needed in a small dingy work room at the mill.

J. D. worked on the swing shift at the mill, in a building constructed long before there was any way of clearing away the specks of cotton dust that constantly floated through the air, coating everything and everybody with a white mantle. Not surprisingly, the workers in these obsolete mills were called 'lint heads.' This might have been considered a derogatory term, but was not, because it was applied to those who overcame considerable difficulty in producing viable products with outmoded equipment; and so to be called a 'lint head,' was a kind of compliment.

I finally found out why J. D. was not happy with his wife's visit, when I met with him over a couple of beers at the Post Exchange. After some hesitation, he explained his lack of enthusiasm for the visitation. Back at the mill, he worked the night shift, and his wife, the day shift. This gave him the opportunity to engage in an affair with an attractive gal who worked the same shift as he did. The affair took place when they both were off shift, and his wife was working; and this had gone on for some time. But when J. D. was inducted into the army, and not there to keep things under control, his girlfriend just had to confide with her best friend about the romance she had been experiencing. Unfortunately, as is usually the way, the best friend needed to confide with someone else about the relationship between J. D. and her friend. From that friend to that friend's best friend and on to another friend, the news of the affair circulated through the mill community, until it reached J. D.'s wife - and that was why she was coming to Fort Rucker. J. D. was not looking forward to the visit, which, as he feared, turned out to be a disaster, and the end of his marriage.

Later, the 'old men' were given their early discharges; even though J. D. was perfectly able to continue to function as an artificer, he was one of the first to obtain the required job in a defense plant and depart from the 98th Division. But not back to his small town - there was no point returning to a place where everyone knew about his misadventure. This time he was going to try and make his way to a different place, a defense plant in the Atlanta area, where he could start anew.

Finally, only two of the 'old men' were left. Although private Lindsley, who had walked a trap line for many years, was forty-five, he had no intention of leaving. He liked the infantry. "Geeze," he said, "three squares a day and light duty?," referring to the rigorous Army training that many younger men found exhausting. Infantry duty was to Lindsley a breeze, not much more difficult than the many miles of trap lines he walked every day of his life, rain or shine. And to cap it all off, he was getting paid for this "easy life!" The other 'old man' was Abdul Bashir. Abdul, also forty-five, from Bengal India, had been in the British Army there, but was drafted when he emigrated to New York. The discipline and army training learned in the British Army served him well, and although he was slight of stature, he never lagged. He had a few peculiarities: his salutes were British style, and he called the platoon commander "Leftenant." But otherwise, he fit in well enough, even though some of his customs, combined with some language problems, made him somewhat of an outsider. Surprisingly, he had not taken advantage of the chance to leave the Army; although, with most of the other 'old men' gone, he was pretty much by himself.

Finally, my curiosity about his rather lonely existence caused me to ask him why he was still around. "Sergeant," he said, speaking in his high- pitched Anglo-Indian accent, "you have to write a letter to a defense plant, to get them to ask you to work for them." "I can write," he said, "but not in English." To correct this problem, I wrote a letter on his behalf to the Ford Motor Company in Dearborn, Michigan. Abdul signed it, and it was sent off. Soon there was a reply, and Abdul was off to his new life in a factory in the mid-west, where in all probability there were not too many fellow Indians, let alone Bengalis. Eventually, I received a letter from Abdul, but I never did learn what it said. The letter with beautifully decorative handwriting, in a strange language, was a work of art. Unfortunately, there was no way of translating his letter in the Bengali language, to English. So there was no way to find out about his new life in a factory in the American mid-west. And I often wondered about his existence, in what must have been for him a strange environment.

That left just Lindsley; and in spite of the army life he admired, he eventually succumbed to what he saw as a serious flaw in army life. On a recent Saturday night, he had gotten drunk at the Post Exchange. And while it takes a lot of serious drinking to get high on the low, 3.2%

alcohol content of PX beer, he had become boisterous and combative. It took three M.P.s to take him down and lug him off to the holding tank.

That episode changed Lindsley's opinion of army life. He was vehement about his change of opinion: "Any place where a man can't get drunk on Saturday night," he maintained, "is no place to be." And so, he sent off his letter, got one in reply, and was gone.

Shortly, able-bodied replacements for the old men, and the other rejects, began to arrive at Ft. Rucker. These were not all new recruits, many of them coming from somewhat cushy military assignments. Early in the war, the government, realizing that college training was important to the development of a professional military effort, had introduced the Army Student Training Program (ASTP), and promising young recruits from all over the Army were selected to be part of this program. Many of them came from fairly routine sections of military life, where their duties often had not been onerous. Their new assignments at ASTP usually consisted of classes in colleges, where they were remarked upon with favor by their non-military female classmates.

Now, because the high casualty rate in the Pacific war indicated that the need for replacement troops would be extreme, the War Department decided to abolish ASTP, and transferred these cosseted students into a humiliating new life as infantry soldiers. Naturally, the former students who quite often generally had cushy jobs in the military, even before ASTP, did not like this one bit, and they often made their unhappiness known, bitching a lot to anyone who would listen. The usual response given was the typical army suggestion for problems - that they should go to the chaplain's office and express their problems there. In other words, "forget it."

New for them and normal for the rest of us, our grueling training went on. Fortunately, it was winter. The climate in southern Alabama, which in the summer could be brutal, was now moderate; and things were easier with water discipline left back in the 2nd Army, with General Lear and new recruits. Those of us remaining with the 98th, hardened by the many months of intensive, rigorous training, could handle this more relaxed army life with somewhat lessened aversion. As our intense training continued, it was obvious, although seldom mentioned, that it would not be long before we would be in combat somewhere. It was the elephant in the parlor, though seldom mentioned, that was always there.

## Working with the Artillery

FROM TIME TO TIME, beginning shortly after my enlistment, I was selected for several different special assignments with units that had different procedures from the standard practices of the infantry. One of these special projects involved a coordination program with a unit of our divisional artillery and learning how they operated. As the main function of the 98th Division's artillery was to provide support for the frontline infantry, mutual understanding of how we would talk to each other on the battlefield was of paramount importance. My group, the infantry, had been in the beginning, that we were the ones who just walked; and our support facilities, the divisional artillery, mostly rode in trucks and caissons, except for those who were pilots of the 98th artillery observation planes.

The Army Air Corps did not see coordinating with the infantry as their primary mission; and so, forward-firing targets were relayed to the artillery by small Piper Cub-type observation airplanes piloted not by officers, but by the artillery sergeants. I did not deal with any of the pilots. My counterpart in the artillery was a technical sergeant who was very knowledgeable about how things worked. This might have been because he was not only a veteran of another war, the Spanish Civil War, but had been an officer in that war. Although he was an American citizen, he was anti-fascist, and had volunteered for that war.

He had become an officer in the Abraham Lincoln Brigade, a rather large group of idealistic young men who had gone to Spain to battle the Fascist army of Generalissimo Franco. Franco won the war with the help of the German Nazis and the Italian Fascists, but fortunately, most of the Abraham Lincoln Brigade managed to make their way back to America. As soon as America became part of WWII, my artillery associate joined the American Army, expecting that with his wartime experience, he would be welcomed as an officer in the Army. However, in the War Department, there was a fear that Abraham Lincoln Brigade veterans might be too radical to be officers in the U.S. Army. So, although he was rejected as an officer, he became a non-commissioned officer, with the rating of Technical Sergeant.

He was particularly helpful by explaining to me the various ways of working between the artillery and frontline troops, a coordination that would be needed when the invasion of Japan was underway. I felt that I was fortunate to have known and worked with a technical expert, and that at least I now had some understanding of the problems involved in coordinating communications between two very different branches of the service.

## In the Deep South, A Statue to a Bug

MEANWHILE, CHANGES WERE TAKING place on the base, with the old soldiers fading away, along with most of the physically unfit. Along with new recruits, and new officers coming in as well, the 98th Division was drastically renovated. Within the confines of Fort Rucker, much energy was created not only by the changes, but by things going on within the base. A giant military establishment, Rucker boasted a continuous round of activities, including all sorts of games, amusements and various entertainments. While helpful to morale, these on-base activities did not change the fact that there was not much going on away from the camp.

Fort Rucker, like most military bases in the South, was named for a Confederate officer, Edmund W. Rucker. He was a Confederate hero, who had been given the honorary title of General; and in southern Alabama, the honorary title of General in the Confederacy was meaningful. Memories of the Confederacy flourished in the area, and the local residents, after years of inundations of military personnel, were not partial to soldiers - particularly those from the North.

The civilian area surrounding Camp Rucker - the small towns, and even the somewhat larger city, Dothan - did not have much in the way of interest for those of us on weekend passes and other off-duty free time. The exception was Enterprise, a smallish town that sported a highly unusual attraction, an 18-foot-tall statue of a bug, the boll weevil.

Strangely enough, this unusual attraction was not in honor of the bug; but rather because its depredations had forced the local farmers to stop trying to grow cotton, an increasingly worthless crop. Over time, cotton had depleted the soil in southern Alabama to such an extent that growing it was no longer profitable. The whole area was facing economic collapse with no prospect of improvement, when a bug, the boll weevil, came up from Mexico and completed the job, destroying what was left of the spindly, sickly cotton plants.

At this point in time, there was not much hope of a replacement crop, and despair was just about everywhere - except at Tuskegee Institute, a noted agricultural college created after the Civil War to provide higher education for African Americans. There, a prominent member of the Tuskegee faculty, George Washington Carver, born into slavery, was nationally known for his work on solutions to southern agricultural problems, including boosting growth and end-use adoptions of his promotions of new uses for sweet potatoes and soybeans. He also was recognized as a champion of soil retention, and on the promotion of agricultural education.

He had been working on peanuts as a viable crop for southern Alabama, not only as a profitable farm product, but also as a conditioner of the soil. Because of Carver's research, the value of peanuts was quickly recognized, and this easy-to-raise legume not only restored the soil but proved to be a profitable agricultural product. Extensively planted, peanuts created an economic bonanza for southern Alabama.

Something needed to be done to celebrate the revival of good fortune; but how to do it? There was no way that recognition of the efforts of an African American scientist at Tuskegee could be publicized in southern Alabama. So, instead of identifying George Washington Carver as the scientific force behind the success, a statue was created to honor the other agent of change, the boll weevil.

And so, a statue of the bug that caused the demise of the cotton crop became the symbol of the economic resurrection of southern Alabama agriculture; at eighteen feet tall, it was a noted landmark. Without much else going on in Enterprise, the otherwise backwater county seat, the prominent statue of a bug, the boll weevil, became a sight for tourists and off-duty soldiers to marvel at.

By spring, our waiting period at Camp Rucker was over. Evidently, the Army felt that the reconstruction of the 98th Division in terms of the restoration of troop strength in numbers, and the training of new recruits and the replacement of commissioned and non-commissioned officers, was far enough along for the Division to be shipped overseas.

# Part Three
# Trip to Hawaii

# Trip to Hawaii

After our sojourn in southern Alabama, orders that came through for our next assignment indicated that we were on our way overseas, and to whatever fate awaited us there. There was no word given about where we were heading; but when additional clothing and equipment came through marked "tropical," it was pretty evident that we were destined for the Pacific Theatre. On our trip to an unknown port of embarkation, our train traversed spring vegetation while we were in the South; but as we traveled northwest, winter was still in evidence. That our port of embarkation turned out to be Seattle was not a surprise; and without much ado, we shortly embarked.

Our ship was a former luxury liner, the *Lurline*, the flagship of the Matson Line, now converted to a troop ship. By some stroke of luck, my squad was berthed in a second-class cabin, instead of being out on a half- enclosed companionway along with the rest of I Company. The cabin, designed for two, was now home for our squad of twelve and all of our equipment. Crowded, to be sure, but still better than being out on the companionway.

Eating was something else again - no semblance of luxury here, as least for us, as we stood in line for hours - in contrast to officers, who strolled in at their leisure to sit at tables graced with linen and silver and served by uniformed waiters. We knew this because we passed by a window that looked in on what must have been the first-class dining room, as we descended to the bowels of the ship - there to partake of a rather pressured and somewhat unappetizing meal, served by mess personnel in rather dismal surroundings. We stood as we ate at long tables, moving along as we ate. You had to eat quickly, or you would be pushed off the end of the table before you were finished. However, the quality of the food was such that we were not tempted to linger.

On the troop ship, we spent a lot of time in lines, even though only two meals were served each day. Even the showers had a line. They were saltwater showers, which made bathing a different experience, as we had not been issued saltwater soap. When I tried to wash my hair with ordinary soap, it didn't lather up in the saltwater. Instead, it caked up and my hair became tangled and bristly, my scalp resembling something like a Brillo pad until some saltwater soap was obtained.

The trip itself was relaxing; the Pacific was pacific, and the weather was beautiful. There was always a place to sit outside; or you could stay inside and join one of the games of chance that were the passion for many. I chose to sit outside; and my favorite spot was the paravane, a device that stuck somewhat out from the deck to hold a cable that could be used to cut mines, in case we should ever deploy into a mine field. There were plenty of books available, and the paravane was a cool place to sit and read. And so the time passed until the much anticipated islands of Hawaii came into view.

My first sight of Hawaii was of just a smudge on the horizon. In time, the smudge became a cloud, and as my expectations soared, a mountain grew, sharp-shaped against the fluffy clouds. Finally, the high plateaus came into sight, and with them, the first sign of human habitation: cultivated fields, roads and little villages. As the ship turned toward port, the lowlands appeared - great tracts of greenery broken by small harbors, buildings, boats and palm-fringed beaches. The gradual revelation of the island, and its extraordinary features as it rose from the empty ocean, was a rather magical process.

The spell of old Hawaii was broken by our arrival at our temporary destination, the port of Honolulu. It was busy with all of the paraphernalia of a modern seaport: the docks loaded with materiel destined for the far-flung war and numerous cranes, winches, and small cargo craft, all at work. Boats were everywhere: warships, merchant vessels and harbor craft – all seemingly in motion - coming, going, loading, unloading, at anchor, or tied to piers, all part of the enormous effort needed to supply the war in the Pacific. Much as I would have liked to have seen more of the port and explore the fabled city of Honolulu, our stay was short; and our sea voyage resumed without a chance to go ashore.

Once back at sea, it was announced that our destination was Kauai, a neighboring island to Oahu. It was known as a garden island, we were told, a quiet place as opposed to lively Honolulu. This was a thought that did not thrill us, as it seemed from some rumors that we were going to be stuck in a place where nothing much was happening, while the division continued its reorganization. What would happen after that was the unanswered question, as our projected eventual entry into the Pacific War was grist for the rumor mills that daily ground out cock-and-bull stories for the gullible.

## LIFE ON KAUAI

WITHOUT STOPPING IN OAHU, we had changed ships in Honolulu Harbor to a smaller inter-island steamer that transported us to Kauai, our destination for training and guard duty. On the way to Kauai came scenic headlands: volcanic cliffs, great walls, turrets and pinnacles of lava rising from the sea. Huge waves that had rolled halfway around the world without impediment were now crashing with great force against the cliffs, sending a wild spray high in the air, creating a magical mist that gave the wildly sculptured headlands a fairy tale look.

Once past the rugged seaside cliffs, Kauai changed radically into what was called the 'Garden Island.' And it was not misnamed; with rich soil, abundant flowers, large sugar plantations - and many pineapple and banana farms - all interspersed by small seaside towns. After the hubbub of Honolulu, Kauai was almost like a visit to Brigadoon. During the California gold rush, the '49ers sent their children to these islands for their education, and Kauai was mostly unchanged from those early days. With few visitors, not even Hawaiians from other islands, Kauai maintained an eerie quiet that gave you the feeling that you were living in a place where time, if not stopped, had slowed way down. A feeling of peace emanated from the landscape: the historic public buildings, the early New England-style churches, and the people themselves who seemed to be living in another era. Everything about Kauai was fascinating. It was like living in a museum, even with our military duties, that our time there seemed to have a sort of other-worldly quality. It was a peaceful paradise, lacking only Eve.

Before inter-island air travel was instituted, not many went to Kauai. The sea voyage was too rough for most to make the effort, which is why there was little of any evidence of facilities of any kind for tourists. Because so few people came, the only housing for visitors, an inn which had rooms for only 20 people, was seldom fully occupied - this, in spite of the incredible vistas. The beaches were spectacular, as was the scenery in general, including Waimea Canyon, with views there rivaling those of Arizona's Grand Canyon. The towns, Hanapepe, Nawiliwili, and Lihue, were sleepy places, where nothing had changed in years. Small stores, run mostly by people of Japanese ancestry, sold basic goods. Some color and excitement came from those who worked at the sugar plantations. Mostly of Philippine extraction, they maintained the village life of their former islands, with colorful fiestas that were enlivened by local moonshine and exciting cock fights.

Native Hawaiians, who had never adjusted to organized farming, continued their Polynesian way of life, gathering taro for poi, and hunting wild pigs in the jungle. They employed hand-held, homemade nets to catch small fish off the reefs, and kept harpoons handy to spear larger fish and occasional sea turtles. Other than the friendly sounds of an occasional family get-together, in the form of a Hawaiian party known as a luau, quiet settled comfortably on the land and the people.

While much of Kauai consisted of rugged mountains, sugarcane was grown whenever the land was flat enough. The Robinson family were the primary plantation owners, and they owned most of the arable land on Kauai, with others leasing from them. The nearby island of Niihau, also owned by the Robinsons, was a restricted island, where only full-blooded Hawaiians were allowed to reside.

Everything about Kauai was fascinating. It was like living in a museum, so even our military duties had a sort of other-worldly quality about them. Kauai was a peaceful paradise. It was like Eden, before the creation of Eve.

## Special Duty in Paradise

SOON AFTER LANDING IN Kauai, the 98th was divided into groups that alternated between guard duty and jungle training. Both of these activities resulted in an appreciation of the spectacular environment surrounding us. In my first case of special duty in Kauai, my squad of twelve was placed on detached detail, and we had the good fortune to be given an incredible assignment: we were ordered to maintain a lookout for airplanes from an observation post on the spectacular cliffs at Na Pali. We were thus provided with a particular paradise, because on Kauai, the usual island-circling road only goes about three-quarters of the way around the island; because at Haenea, the massive and beautiful Na Pali cliffs rise directly from the sea, preventing any sort of continuation of the road.

Our base for the lookout was a small, pleasant cottage by the sea, the only dwelling in this enchanted area. It was on a spit of land at the base of the Na Pali cliffs and housed my squad of twelve G.I.s. Food for the outpost was brought in from company headquarters down the road; otherwise we were completely on our own. The cottage overlooked the ocean, our own beach, and a tidal pool. The pool was made possible by a massive coral reef that enclosed it, except for a small break in the reef that allowed the sea to come in and go out.

The reason for our being in this paradise was that an airplane-spotting tower located high on a nearby steep and rugged cliff had to be manned night and day. Manning the tower was to be the main responsibility of our detached assignment - that is, if you would want to call being in paradise a responsibility. The tower had not been there before the attack on Pearl Harbor, but now was always manned by a crew of two. Telephones connected it to the base, so that a flight of planes, or even an individual plane, could be reported.

We were given the rules and regulations by the departing crew as they reluctantly turned paradise over to us. We gained, among other things, a delightful tidal pool for swimming, and an intriguing coral reef that was always an adventure to contemplate. The path up the steep cliff from our base camp to the observation post was very challenging, requiring both hands and feet to make the climb, as it was difficult even in broad daylight. Those on night duty ate early and made their way up before dark. The view on the way up or down was fantastic, but a slip or misstep could send you falling hundreds of feet to the ocean below. The two-person lookout changed twice a day, once in the morning and again in the evening, while it was still light enough to see your hand- and footholds.

The duty arrangements seemed to be working, except for one evening, when one of the spotters, Olin Lynn, a backcountry mountain man from West Virginia, failed to show up when it was time for the evening shift to take over; and so the other member of the lookout team went up alone. Private Lynn had not appeared before I turned in, although I stayed on duty long after dark, as I was concerned that his missing his tour of duty at the tower would put him in deep trouble. I finally gave up and endured a troubled sleep. However, the next morning, when the night shift was relieved, I was astonished to see Private Lynn come down from the tower with the other member of the night crew. He was somewhat disheveled, and out of uniform, as he was barefoot; but obviously, somehow he had been able to make the dangerous climb in the dark to the observation post in time for his tour of duty.

It turned out that he had done just what he would have done back home: take off his shoes and socks and climb barefoot up the perilous path in the dark. The fact that he was well juiced did not impair him; that was just like back home, too. For me, it was a difficult enough climb during the day, as the path was very steep, and for most of the way, you were just on the edge of a sheer drop-off. All of this made the climb a hazardous task for me, even in the daylight; but then, I was not from the mountains of Appalachia.

Setting up and working out the details concerning the manning of the outpost did not take up much of my day, but investigating the mysterious and possibly dangerous reef that enclosed the pool was always an adventure to look forward to. One time, while exploring a section of the reef a little further on from our base, I saw a local fisherman harpoon a giant sea turtle. Hauling it in and turning it over, he butchered it on the spot. As this happened when I stopped to talk to the fisherman, I could not avoid seeing the slaughter. I knew that the meat was an appreciated delicacy to the native Hawaiians, but it was an unsettling sight. Another time, walking on the rough and slippery reef at low tide, I noticed a giant moray eel, his head sticking out of a hole in the coral. As I watched, I saw that it was observing a spiny sea urchin some distance away. When the urchin turned so that its unprotected bottom was exposed, the eel flew through the air, and, bypassing the spines, hit the exposed bottom, smashing into it, killing the urchin instantly. After seeing how large and fast the eel was, I decided to curtail my explorations of that part of the reef.

The base camp was on the edge of a small beach that was bordered on one side by a jungle, and on the other sides by the reef. It was a small, delightful swimming hole that was quite private, stretching from the beach to the reef that ran parallel to the beach. A small opening in the reef caused a narrow passage that led out to open sea. This opening created a current that at certain times created a tidal rush. But most of the time, the water in our swimming hole was calm, and swimming there was very enjoyable. For those of us who could swim, we made floats out of cotton mattress covers. It was strange; but if you soaked the mattress cover and held it up to the wind and quickly tied up the end, you had a float that you could ride on, which most of us did at one time or another.

The reef was an interesting place to observe sea life amidst the coral, and the strange creatures that constituted life on the reef. One day, I was out on the reef, poking around, when a commotion arose. Tsosie, the Navajo, and Solice, a young Tex/Mex, both non-swimmers, were on a mattress cover float that had become caught in the tidal run and was being carried out to sea through the gap in the reef. There were no boats available for their rescue if they got beyond the reef, and they were almost at that point. Fortunately, several of us out on the reef were close enough to run over to the opening, and at the last moment, grab the mattress cover with its two non-swimmers, and drag it, with them still on it, back to the beach. The rescue had been the non-swimmers' last chance, as the current had floated them close to the open ocean, where the waves were always rough, and rescue dubious. Although we continued to enjoy the tidal pool as long as we were on guard duty there at the base of the Na Pali cliffs, the floating mattress covers, if used at all, were used with caution.

## Island Entertainment

THE QUIET AND UNEVENTFUL "Eve-less Eden" that Kauai was for us, occasionally had a blip in the even tenor of its days. The disruption of this peaceful paradise happened twice - each time, when a USO show paid us a visit. The shows produced by the USO in WWII were a tremendous combination of talent and artistry, including the two that came to our out-of-the-way island. These two shows ranged from classical music to classic comedy. The classical music was a concert by world-famous violinist, Yehudi Menuhin, and the classic comedy was a Hollywood-type variety show, emceed by none other than Bob Hope.

Classical music was not the usual kind of program that most of us basic infantry types were used to, and so there was not a lot of enthusiasm for an announced evening of classical music. Our amalgam of hillbillies, Brooklyn hepsters, plus the old men and boys from everywhere, most of whom had never been to a classical concert, had never heard any classical music, nor were interested in hearing any. The thought of having to listen to what probably would be boring stuff, particularly by a guy with a strange sounding name who played not a fiddle, but a violin, was way out in left field for most of us. But, with nothing else to do, we went anyway. To our amazement, the music was astoundingly beautiful, with magic sounds floating out from the stage. The applause, tepid at first, grew to a roar, as song after sublime song soared out.

When the regularly scheduled program was over, it was announced that encores would be by request from the audience. The requests went on for some time, as we did not want to end the evening. But then, someone in the audience asked for "Flight of the Bumblebee." As this very difficult piece was perceived as a cliché, the maestro shrugged it off and played another piece. Again, the request for "Bumblebee" was called out, but was again shrugged off. The call for "Bumblebee" became even louder, coming now from many voices, most of them from soldiers who probably had never even heard of the old cliché. The clamor from the audience was so strong that the maestro, who had been about to leave the stage, held up his bow for silence; and when the sound died down, began to play the piece that he had been avoiding. It may have been old hat to him, but for the audience it was thrilling; and when the last sound of the music died away, pandemonium ensued. I don't think anyone who was there ever forgot what was, for most of us, our introduction to classical music.

Our introduction to a comedy/variety USO show was another matter. With Bob Hope as the master of ceremonies, it was Hollywood at its best, with scantily clad young ladies singing, dancing and cavorting on stage, with Hope interjecting his witticisms at appropriate times. It was everything a Hollywood variety show should be, and of course we loved it. Always on target, Bob even seemed to be in sync with the G.I.'s life on Kauai, as his aphorism about the "Eve-less Eden" was a paraphrase of a Winston Churchill quote. Bob said, "Never have so many chased so few, for so long, for so little, or for so little success."

While the USO shows were much appreciated, the acknowledgement about Kauai being an "Eve-less" paradise made the next step, a move to Oahu, not that much to be upset about. We knew that ultimately, the reason for the move was because of the soon-to-happen invasion of the Japanese homeland.

Momentous victories by the United States in every battle with the forces of Imperial Japan pointed the way toward the upcoming invasion, and we knew that we would be part of the troops that were being assembled and trained for that crucial conflict. And so, while we knew that the reason that our part of the 98th was being moved from quiet Kauai to lively, bustling Oahu was to prepare us for our part in the momentous invasion. It was a disquieting, but not unanticipated, change of focus.

## Training on a Wet Mountain

OUR TRAINING IN REMOTE jungles, guard duty in restricted areas, along with general maneuvers in places that were off the beaten path, permitted us to see aspects of Hawaii that were not available to most non-islanders. Some experiences, however, we would have willingly skipped, including a training exercise that featured climbing Waialeale, the tallest mountain on Kauai, and the wettest spot on Earth. This mountain, created by an extinct volcano, experiences rain 350 days a year, with a record-breaking average rainfall of 467 inches in one particular year, which means that it rains almost every day-- not just showers, but heavy rainfall. Our mission was that we were to climb this mountain and spend the night on the summit. The military purpose of venturing onto the wettest spot on Earth was never explained to us.

Trucked to the base of the mountain, we slogged upward on foot, encumbered with the usual pack, steel helmet, and rifle. We struggled on, slipping and sliding through damp, mucky jungle trails until, late in the afternoon, we achieved the summit. Giant fern trees loomed among other oversized and sodden vegetation, creating an unworldly, almost underwater, atmosphere. As we squished around the mountaintop plateau, our boots kept us dry for a few steps and then filled with water, making every step waterlogged. There was no dry spot to sit or recline, so we stood in organized discomfort in the fading light. We were issued dry packaged K-rations for supper, which we had never found particularly appealing under normal situations. But cold rations in a dank, drizzling jungle was not a meal to be remembered favorably.

What happened next was not a happy moment, as we were informed that we were to sleep there, on this very wet mountaintop. The watery blankets we spent the night wrapped in did not induce quality sleep; but when you are young and healthy, it seems you can lie down in a pool of water and obtain at least some sleep. In the morning, we wrung out our blankets, packed them up, and more or less slid down the mountainside to much appreciated dry land. Wet and squishy as we were, getting away from Mount Waialeale made us, if not happy, at least glad that the wettest place on earth was behind us. I never did find out the purpose of our upward slog to the summit of this very wet place; or why, on another maneuver, we spent some time in an area, not far from the wettest place, where it had never rained in the history of man.

## Na Pali Valley

THE INVOLVEMENT OF MY squad with paradise did not last. The Third Battalion was given the assignment of executing a bivouac on the other side of the Na Pali cliffs, in the remote and difficult-to-reach Valley of the Lost Tribe. We were expected to join the rest of the regiment; so, for this reason, we turned over our short- term utopia derived from a watchtower high on the cliffs to another lucky contingent and joined the trek through and over the cliffs. In spite of their forbidding aspect, the Na Pali cliffs were not entirely impassable. A path led us to an opening in the cliffs where the waves beating against the base of the cliffs sent spray high in the air, creating spectacular images in this already enchanting scene, and showed us the possibilities ahead.

On the other side of the opening was an eleven-mile-long, narrow and winding path. This rugged path staggered through various narrow passageways and clefts in the cliffs; and in places, we followed the path up and down and along the edges of the cliffs, to reach the almost landlocked and quite inaccessible Valley of the Lost Tribe - a place that could only be reached by this difficult path, or by the sea.

Handicapped by our usual equipment: pack, rifle, helmet and gas mask, it was slow going along the uneven and twisting path. The journey turned out to be an all-day trek, not exactly a march, but more of a struggle. It was worth the effort though, as the very verdant but uninhabited valley, surrounded by steep cliffs on three sides, with the sea on the fourth side, had plunging waterfalls and generally spectacular scenery.

It was interesting to see platforms for long-ago abandoned native houses scattered through the valley. At one time, this lush valley had been intensively cultivated, but now, much underbrush had since grown up. The beach was good, but it was dangerous to swim there, as the undertow could be lethal.

Provisions for our expedition were brought by sea, using landing craft that could come right onto the beach to be unloaded. The landing craft arrived at about the same time we did, so we were able to help unload it, and also help their passenger, suffering from exhaustion, get to the field hospital. He was a rash swimmer who was rescued by the crew of the landing craft while it was coming in for the landing. He was from a Navy group that was temporarily stationed there, and while swimming, had been carried out to sea by the rip tide. It was just pure luck that the crew of the landing craft saw and rescued him before it was too late.

After the rescue, the Navy guys started a small bonfire, and we sat around, some of us drying off, and as usual, told stories. Some of them obviously were mostly invented, but some were insider backstories that were explanations of some of the strange events that were happening around us. One of the Navy people, a tall rangy guy, whom they called "Chief," always seemed to have a cigarette in hand or in his mouth, and seemed to know a lot about the history of the islands and what was happening in the war. As I had been curious about what could cause a small country like Japan dare to take on the much larger United States, I asked Chief if he knew how it all happened.

He was happy to tell us how it all began, and seemed to know the full story, as I thought he might. After opening a fresh pack of cigarettes and taking a few puffs, he started with some history. He said that "the U.S. in the early part of the 18th century persuaded Japan, a hermit kingdom, to open up. Before that, under the rule of the Shoguns who ran the place, no one was allowed to enter or leave Japan, and even shipwrecked sailors were not allowed to come ashore there. To keep the population in balance, a primitive form of birth control was practiced, which kept the population stable for centuries. But after the opening of Japan, Christian missionaries visiting Japan preached that birth control was a sin; and so, in the new Japan, which was trying to be like the West, birth control was no longer used."

However, the resulting population explosion prompted the authorities to incorporate the population surge into armies and war factories. With the new military might, Japanese armies conquered much of the Far East, including Formosa, Korea, and China, starting with Manchuria. This prompted the small but highly organized Chinese communist party to leave the coastal area, leaving the Nationalists, under Chiang Kai-shek, to attempt to resist the Japanese takeover of much of China. As the U.S. considered the Nationalists an ally, Chief told us that the U.S. tried to combat the Japanese invasion of China, at first supporting military aid to Chiang's armies and also by diplomatic pressures. Neither effort worked, so harsher measures were applied. These consisted of cutting off the Japanese from supplies of oil and rubber from the Dutch East Indies. In Europe, Holland was under the yoke of Nazi Germany, but the Dutch East Indies were sort of under the protection of the United States.

As the Japanese military could not function without the oil and rubber from the Dutch East Indies, they were faced with a major problem. If they tried to take over the Dutch islands in the Pacific, they would be confronted by the powerful American navy. But, while Japan had to do something to keep their military machine from grinding to a halt, it seemed to be an unsolvable situation for them, until Admiral Yamamoto, an up-and-coming officer in the Japanese navy, came up with a radical plan. This was to destroy the American fleet at its home base in Hawaii. With the American fleet out of the way, and difficult to replace, Yamamoto claimed that the Americans would lose interest in the war and agree to a compromise settlement with the Japanese government.

The first part of Yamamoto's plan had worked to perfection for the Japanese, but not the second part. We did not need for Chief to tell us more, as it was apparent. The destruction of the American fleet energized the United States. Even with the nearby losses in the Pacific, we did not lose interest in the war, and that is why we were now preparing for the invasion of Japan. The group broke up soon after Chief explained the war to us, and the Navy guys went back to their landing craft and were off to some other challenge.

This revelation of how the Pacific war began gave us a lot to think about, as we went about our few days allotted to us in the valley. There was not a lot to keep us busy, outside of communing with nature, exploring, and goat hunting. These goats were not native animals, but farm animals that had escaped several generations ago, and become truly wild. Omnivorous eaters, they were destroying the vegetation, and so were legitimate targets. We saw some, but they were too far away to zero in on. In spite of the splendid beaches, we did not do much swimming -- the earlier last-minute rescue was a warning that we observed.

After three days, we straggled back over the rugged path to our next assignment, a session of jungle training exercises that were combined with researching the defensive potential of the area. Among the outstanding scenic vistas that might be targets, Hanalei Bay was a possibility for us to look into. The bay was calm, quiet and absolutely beautiful, with only a small fisherman's wharf to disturb the serenity. It was a well-known and much-loved place, with its own song celebrating this charming spot; and it was obvious that, post war, this would be developed, and I am glad I saw Hanalei Bay before that happened. In exploring the area, we concluded that, because of its extended expanse, Hanalei Bay would be difficult to defend against an organized attack.

We moved on. Between our lookout at Haenea and Hanalei Bay, the beautiful Kaupea and Lumahai beaches were mostly deserted in those days, but were later made famous by being the place where the movie "South Pacific" was filmed. Another remarkable local attraction was a hidden pool, high on the Pali cliffs, whose water was a deep blue color. That was to be found, we were told, in a cave that was located quite high in one of the cliffs. Following directions from a local authority, we found the cave by climbing part way up a high hill and then down a murky path to the underground pool, that was, as they said it would be, filled with deep blue water. We tried swimming in the dimly lit pool among stalactites that hung down from the ceiling almost to the water. Swimming in a dim light in bright blue water amidst multiple stalactites was an eerie experience. As the water seemed to be moving toward the back of the cave, I tried to see where or what the water was moving towards; but the stalactites that were closer together in the back of the cave kept me from swimming all the way to the rear, although a current seemed to be flowing in that direction.

Back in the bright outside, in an area that was not far from Hanalei, there was a traditional rice farm run by a Chinese family. As they were fairly recent immigrants, the farm was operated much as farms probably still are in China. This local farm had a flooded field where the rice grew, with a tall pole, like a telephone pole, in the center of the field. The pole had a perch-like seat on the top that was shaded from the sun. Radiating from the perch in all directions were strings with tin cans attached at intervals. The cans were partially filled with loose stones that could produce rattling sounds when you pulled on the string. This was their bird protection device. A person, usually a child, would sit up there on the shaded perch, watching for birds. When one alighted anywhere on the paddy, the watcher on the perch would pull on the string that led to where the bird had landed, thus shaking the can on the string there, making the stones rattle, frightening the bird away. It was a way of protecting a rice paddy from hungry birds that probably had been used by the Chinese for thousands of years, and now somewhat new to Kauai. But, like everything else there at that time, it was reminiscent of times past.

It was a kind of Eden there in Kauai back then, but like Eden before God created Eve. And while we, the soldiers stationed there, pretty much all agreed that there was nothing like a dame, there weren't any there for us. Eventually, the Eve-less Eden ended for us, as it was decided that Kauai was now safe from the enemy, and we were relocated to lively, up-to-date Oahu to prepare for the attack on the Japanese mainland.

the field. by S/Sgt. Richard Parker

# Part Four: Oahu

## Back Country Oahu

THE WEATHER IN OAHU can usually be described as fabulous. Trade winds blow across the Pacific 50 weeks of the year. This leaves Honolulu with a perfect climate and fair weather most of the year, except for two weeks in September, when the trade winds stop. At that time, Honolulu becomes yet another tropical island with a hot, sticky climate similar to that usually found the rest of the year on the other part of Oahu, known as the Leeward or Eva side. These now uncomfortable few weeks in Honolulu are known as Kona weather.

When the rain-swollen trade winds hit the mountains that stretch across the central part of Oahu, they dump their moisture on the back side, causing Kona's hot, sticky weather that persists most of the time. This damp, drizzly weather encouraged the wildly entangled jungle that we were to use for our advanced training.

When the 98th was moved to Oahu, the 3rd Battalion was quartered at the monster military establishment, Schofield Barracks. And, as our future seemed to be as jungle fighters, most of our military activity was planned for the rain forests, bordering the mountains in the island's center.

Training in the jungle was a fascinating experience, because at least for a while, we pretty much had the entire backcountry to ourselves. Occasionally native Hawaiians would arrive, hoping to pursue their favorite sport, pig hunting; but as we were using live ammunition with our maneuvers, they did not linger.

A little later, we had instructors who had real-time experience, not only in the jungle (as they were native Hawaiians), but had actually been in combat in the jungles of southeast Asia, including Guadalcanal and other tropical battlegrounds. These Hawaiian instructors were under some kind of detention, as a guard or two loitered nearby; but they seemed to enjoy sharing with us their rainforest experiences. As an example, they showed us how to cook rice in bamboo containers that were cut from large green bamboo trees. They were filled with water and rice, with the opening covered with leaves tied down by vines. The containers were then set by cooking fires, but did not burn through, as the bamboo was green. This allowed the water to cook the rice when perched on the edge of a fire. They also taught us other ways of jungle cooking, and considerable other jungle know-how, particularly about military experience, such as how to avoid booby traps and also how to lay them. Most importantly, we learned how to patrol in dense foliage and how to fire your rifle jungle-style: instantly from the hip (snap- shooting, it was called), and how to use a garrote at night for infiltration killing.

It was not until later that we learned that our instructors were, in fact, military prisoners. Natives of the Hawaiian Islands, these soldiers had been brought here from the South Pacific as part of the increased force that would be needed for the invasion of Japan, but had not been given leave when they arrived back home. So they took off en masse, spent time with their families, and then reported to the authorities. They were loosely guarded and when they finished training us for the jungle, they went back to the stockade. This was not the fearsome stockade at Schofield, but a local enclosure, more like a family gathering.

For the most part, we had backcountry Oahu, and the fantastic jungle landscape, to ourselves. We lived in a silent world that was damp and drizzly, and at the same time, beautiful and mysterious - a place that very few local people, and no tourists, ever visited. Before the sessions with the Hawaiian G.I.s, the dank, lush backcountry was somehow sinister. But now, with our new understanding of how to live with the jungle (including information about edible products that were just waiting for us to enjoy) it was not so fearsome. This appreciation of this environment included the availability of wild fruits like guava and mango. You could smell the mango trees from a long distance, as the fruit seemed to ripen continually, and fall to the ground in quantities where the smell of the ripening fruit filled the air. Because of the attractive smell, it was logical to direct our patrol to a mango grove, where it was a pleasant indulgence to enjoy a ripe mango, on the hot, sticky kind of day often encountered on the Kona side. We also did long-distance patrolling at higher elevations, where the jungle was much sparser, and the air much clearer, and everything looked different. From our patrols and explorations in the rainforests of Oahu, we gained experience and inspiration about surviving life in the jungle.

This valuable information was soon no longer needed, because everything had changed. We were told that the jungle was out, and the invasion of Japan was the new objective. So now, all our training had to be adjusted to "invasion concepts." And with that came something else: the realization that finally, we were going to be part of a long-delayed combat involvement. That also presented the distinct possibility of being a casualty, as the casualty rate among the frontline Marines and infantry was very high in the Pacific theatre. Although this was seldom discussed, it now became, for each of us, an unseen and unremarked on, but always there, *the elephant in the parlor*.

## Fort DeRussy

AFTER MOVING FROM KAUAI, out next post was Fort DeRussy, a military post adjacent to Waikiki beach. With its milling crowds and a kind of holiday spirit prevailing in spite of the war, Waikiki was a far cry from the silent jungle where we had been living, over on the Kona side of the island. The only difficulty was that, in spite of the crowds on hand at the beach, most of the time, we were expected to act as if the war was still nearby. At DeRussy, we handled giant searchlights that could sweep the windward side of the island at night. And, we manned pillboxes situated on the beach, twice a day. At the time of the Battle of Midway, where the course of the war hung in the balance, the American troops assigned to the beach defenses were told, "There will be no surrender." Fortunately, we had cracked the Japanese code; and because of this, the U.S. was able to defeat the Japanese at Midway, relieving the Hawaiian islands of the threat of invasion.

Now, these rather passé defenses were manned at two intervals, dawn and dusk. The Japanese usually attacked by plane or boat at dawn or sunset, with the sun behind them, as it would be difficult to focus on them with the sun in our eyes. So, twice a day we lugged our heavy machine guns out to the pillboxes on nearby Waikiki, set them up, ready to fire, and waited for an hour before dismantling them. At dawn, we were alone on the beach with our death-dealing machine guns waiting for the sun to rise. It was lonely and beautiful then, and it was strangely like our airplane spotting outpost on Kauai . But, like Kauai, although nothing ever happened, or was expected to happen, at dawn the peace of the tropical moment was profound. But, in the evening it was another story. With the beach thronged with strollers and bathers, we felt pretty silly rolling out our heavy machine guns to the pillboxes amidst the beach crowds.

But, by the time we were assigned to DeRussy, the war was far away; and eventually, the guarding of the Hawaiian islands was left to the Navy and the Air Corps. The pillboxes at Waikiki and elsewhere on the Hawaiian islands were dismantled, and we were moved to a tent city on the other side of the Pali Pass from Honolulu, at Kaneohe, where we were quartered while we trained for the invasion.

## From Waikiki to Tent City

THE RATHER LUDICROUS SPECTACLE of soldiers rushing through the crowds of beachgoers as the sun at Waikiki was setting, dragging heavy machine guns over to pillboxes (and setting them up there amidst the hubbub of relaxed onlookers) was a travesty that was about to end. Ever since the lack of response to the attack on Pearl Harbor, defensive positions had been set up all over the Hawaiian Islands, with machine gun alerts twice a day. But now, the shoe was on the other foot. As the war was now approaching mainland Japan, it was Imperial Japan's turn to be concerned about assaults on their homeland.

For us, guard duty in the Hawaiian islands was in the process of being done away with, and the emphasis now was on invasion tactics, carried out from a different campsite. Leaving Ft. DeRussy, our military post near Waikiki , we were sent to join the rest of the 98th Division at another site, Kailua, a remote tent city situated on scenic Kaneohe Bay. On the other side of the Pali pass from Honolulu, our new home, Kailua, was a picturesque, semi-isolated but scenic valley surrounded on three sides by high hills, the fourth side being the ocean at Kaneohe. It was a natural spot to practice invasion techniques.

Getting to and from Honolulu was not that difficult, as city busses used the narrow road that ran up and over the Pali pass. The road on the way up skirted choice residential neighborhoods, giving us a chance to see how the affluent of Hawaii lived, as we passed beautifully landscaped grounds that blended together seamlessly, forming park-like areas studded with luxurious homes. The settings often featured a kind of open-air tropical architecture, where one of the four walls of the living room was omitted. This missing wall allowed the usually pleasant outdoor weather to become part of the outside/inside living arrangements.

These scenes of life in a tropical paradise ended when the peak of the Pali pass was reached, and the leisurely ascent became a hair-raising downward spiral, as the narrow, twisting two-lane highway snaked erratically through the rugged landscape to Kailua and Tent City. On this highway, seemingly out-of-control Kanaka bus drivers wove their way relentlessly down the narrow and winding Pali road, narrowly avoiding whatever traffic dared to be on "their" road. Surprisingly, there were no reports of serious collisions; but accounts of near accidents were often described by shaken passengers. On the other hand, this view of the other side of the Pali pass gave us a look at the primitive side of the Hawaiian landscape.

At Tent City, we had just started pillbox assault training when the day was disquieted by the unsettling news of the death of President Roosevelt. On that evening, April 15, 1945, we were assembled at the parade grounds by ranks and files, as a chaplain spoke a few words of immutable sadness. We stood at attention as our regimental commander, Colonel Lowery, ordered the American flag lowered to half-mast to mark the passing of the President of the United States. As the flag was lowered, the sound of taps filled the still evening air. As the last notes faded away, an echo of taps, sounded by a bugler stationed on a nearby hill, made this a moment of unforgettable memory, as darkness slowly descended on the scene.

The surrounding darkness was not just from night falling. As we broke ranks and ambled back to our tents, we knew that things could now be very different. President Roosevelt had been in full command of the American military, but what came next was anyone's conjecture. Most of us knew little or nothing about our new leader, Harry S. Truman. But, the fact that he had served in the American Army in WWI was, we thought, a good sign. However, what we thought about the leadership qualities of the President of the United States had little relation to our day-to-day issues. We were so far removed from the seat of power that what we thought about what was occurring in Washington made little difference to anyone.

## Navy Special Duty, Oahu

WITH THE KNOWLEDGE OF the imminent invasion of Japan closing in on me, and when I was given another special assignment while still in Oahu, I was mindful of the need for close collaboration between our assault training tactics and the Navy units which would be manning the landing craft. It involved gaining information about the techniques to be used - and I did this by working with Navy mechanics, technicians, and others in the know. From them I learned a lot about invasion tactics and also considerable scuttlebutt about the politics and history of the Pearl Harbor military establishment and its organization.

In addition to what I learned from the special duty assignments, while growing up I was privileged to have a family background that gave me some understanding of the cultures of the Far East. My grandfather was a ship's surgeon on one of the freighters of the Blue Funnel Line that sailed from Vancouver to the Orient, from 1914 to 1929. He was most knowledgeable about the Far East, and from him I gained an appreciation of the mysterious Orient.

One of the groups I worked with at Pearl Harbor was headed by Lieutenant (later Captain) Herbert Rommel. His situation was unusual, in that he was a mustang, meaning that he had risen through the ranks to become an officer. Perhaps because of that, his group was easy to work with. I learned a lot from them - not only about invasion tactics, but also about some of the events that took place at Pearl during the attack by the Japanese and soon after, including some that are still not that well known.

In fact, many of the conversations with military technicians about what had actually happened at Pearl were kind of out there in left field, and not always in line with official pronouncements. But, as they were the opinions and observations of the mechanics and others who were there when things happened, what they said and the way they said it sounded like they knew what they were talking about. Most of what they related was not often anything that could be discussed openly, as some important events that were integral components to the war in the Pacific had to remain secret during the war, and in several cases, even after the end of hostilities.

For instance, it is not that well known that the first shot in the war with Japan was fired by the United States; or how far-reaching, but not immediately apparent, were the results of this first shot. This extraordinary episode took place at Pearl Harbor on the morning of December 7th, 1941 - a day which, up to this event, had promised to be the usual boring Sunday morning for those on duty at the base. The military establishment at Pearl was in its usual relaxed Sunday mode, with officers and enlisted men enjoying golf, attending church, or spending time on nearby Waikiki Beach. The only military force in action was the American carrier force, which was away on maneuvers.

This typical Sunday morning relaxation prevailed in spite of numerous warnings from the War Department of ominous activities by the military forces of Imperial Japan. However, in spite of the warnings, no precautions were observed at Pearl Harbor, America's most important Navy base. It was generally assumed by the military management in charge at Pearl that any hostile activity would take place in the Far East, and that the powerful American fleet based at Pearl Harbor would be able to cope with any contingency.

The lack of defensive preparation by the United States was in spite of the knowledge of Japan's surprise attack on the Russian fleet in 1904. While still at peace, a Japanese naval force attacked and sank most of the Russian fleet at anchor in Port Arthur, a Russian enclave in China. The Russians then sent their Baltic fleet on a long and fruitless voyage to China; on arriving in the far east, the Japanese promptly sank that fleet. J.N. Westwood, in his book on the Russo-Japanese war, summed it up: "The ultimate effect of this surprise attack before the declaration of war was well appreciated by subsequent generations of Japanese staff officers. In this sense, Port Arthur can be regarded as a dress rehearsal for Pearl Harbor."

In 1941, at Pearl Harbor, America was at peace, like Russia in 1904, and in a relaxed mood. Anti-aircraft ammunition remained in its Sunday locked-up position, and the planes at the nearby Army air base were secured behind barricades to protect them from possible sabotage; and were not, in any way, positioned to fly. I heard this recounting of the lack of preparedness in bits and pieces, from various sources (some of it hard to believe) and it took a while to sort them out. But eventually, over time, all of the various pieces of information came together to create an extraordinary story.

While most of the base was peaceful and quiet on this Sunday morning, the crew of USS Ward, the Navy ship guarding the entrance to Pearl Harbor, remained alert and responsive to their duty, which was to monitor all traffic in and out of this important American base. It was an iron-clad rule that to insure immediate recognition, all submarines entering the base had to travel on the surface. So, when the crew of the Ward spotted the conning tower of a mostly underwater submarine attempting to enter Pearl Harbor, they immediately challenged it by radio. When there was no response to their challenge, they did what they were supposed to do: they took action, firing at the only part of the submarine visible, the conning tower. The first shot missed, but the second one scored a hit, causing the submarine to roll over, revealing, as it sank, that it was a midget submarine, a submersible of a type unfamiliar to the U.S. Navy, and in all probability was Japanese.

As required, word of this ominous encounter was immediately forwarded to the military command at Pearl: Admiral Husband Kimmel and his Army counterpart, General Walter Short. Difficult as it is to believe, this incredible information was ignored. There had been many warnings from Washington concerning Japanese military activities in the Pacific. The very aware ambassador to Japan, Joseph Grew, had been advised by a friend, the Peruvian ambassador to Japan, that he had heard that the Japanese were contemplating an attack on a major American base. Grew passed this on, at that time to the authorities at Pearl; and this warning, like all the others, was ignored.

However, sinking a presumed enemy submarine trying to sneak into the United States' most important Navy base should have been, for even the newest recruit, a reason to at least raise an alarm. But, almost an hour passed with nothing disturbing the Sunday calm of the Pearl Harbor establishment, until the tranquility was violently disrupted by the first of the 400 Japanese planes that came roaring in, to begin the attack on the completely unprepared American Pacific Fleet.

The nearby U.S. Army airbase was not spared. Almost all of the American fighter planes were destroyed as they sat on the ground, wing tip to wing tip. This destruction of all of the American planes at Pearl Harbor made the Navy carriers even more important, particularly since all of the planes at Clark Field in the Philippines were also destroyed on the ground.

In addition to the sinking of the Japanese submarine incident, another warning came from an obscure radar post on the outskirts of Oahu. The radar of the observers there showed a large fleet of airplanes approaching. When this was reported to the officer in charge, he dismissed this as most probably a fleet of American planes. He then closed the remote base, sending the observers back to their barracks, and did not bother to report this incident to headquarters.

The assault on Pearl Harbor was not exactly a surprise attack. In addition to the many warnings, the fifty-minute gap between the sinking of a presumably enemy submarine and the arrival of the Japanese planes should have been time enough to unleash myriad defenses. This would have included multiple alarms, forcing the locks of the anti-aircraft ammunition, putting the anti-aircraft weapons on the ships moored in the harbor into action, and using the American pursuit planes from the nearby airports to attack the Japanese bombers as they arrived.

However, while the raid at Pearl Harbor was taking place, there was little military response to the massive attack, except for a few individual sailors who, having rifles, used them; but no Japanese planes were disabled by this scattered and fruitless response.

As a result of the lack of any kind of military defense plan (not even an alarm system), the ignoring of the sinking of a presumed enemy submarine at the entrance to Pearl Harbor, and the ignoring of the large fleet of approaching planes, almost the entire American Pacific fleet was sent to the bottom. Lucky exceptions were our three aircraft carriers, which, with their planes and crews, were away on maneuvers.

The fifty-minute time lapse between the sinking of the Japanese submarine at Pearl Harbor and the Japanese air attack, was so outrageous, that information about it became glossed over, if mentioned at all. That story, if reported, would have been a serious damper on civilian morale, particularly since the only punishment suffered by Admiral Kimmel and General Short was a slight demotion in rank and relegation to desk jobs.

After the war, the families of Kimmel and Short, apparently unaware of what actually happened at Pearl Harbor, petitioned to have the ranks of both officers restored. But when told by Captain Rommel - at that time at the Naval War College - of the fifty-minute gap, they withdrew their petition. To this day, that fifty-minute lapse is not generally known, even though it allowed almost the entire Pacific fleet of the United States Navy to be destroyed in one fell swoop - with the Japanese planes from the raid returning unscathed to their carriers.

With the American navy out of the way, the Japanese war machine turned its attention to the European colonies in the Far East, quickly overcoming Dutch Indonesia and the British colonies wherever they were located. This included the well protected fortress at Singapore, where the British general surrendered a large garrison to a small Japanese force without a fight, thereby condemning the former garrison to inhuman treatment in slave-labor confinement, where many died.

With the total Far Eastern European colonies under their control, the Japanese increased their assault on the American Army in the Philippines.

## On the Positive Side

AT THE BEGINNING OF America's involvement in WWII, the American military establishment was reeling from the disasters at Pearl Harbor and the Philippines. Everyone knew of these attacks, but, because of civilian morale concerns, basic information about the tragic errors that took place at that time had to be kept from the public.

On the other hand, there were military events very much on the positive side for America that also had to be kept secret, this time for strategic reasons. These events were episodes that I learned about while on special duty: military accomplishments that changed the course of the war but had to be kept completely confidential during the war, and some that, even at this late date, are not that well known.

The most important incident of clandestine good fortune came about when the Navy, in spite of being involved in the enormous task of clearing the harbor at Pearl after the attack, but never having had the opportunity to study the concept of miniature submarines, decided to take the time to raise the one that was sunk just outside the entrance to Pearl Harbor by that first shot fired in the Pacific war. While originally ignored by the top brass, that first shot eventually brought about an astounding revelation that, while it had to be kept secret at the time, was of enormous benefit to the United States, a benefit that influenced the outcome of the conflict in the Pacific.

This was because, upon bringing this miniature submarine to the surface and into the lab for study, the Navy struck gold. In addition to valuable information about these novel submersibles, an incredibly important treasure trove was revealed: the super-secret Japanese naval code. This was a document of great importance, not only because it changed the course of the war, but also because it allowed the United States to be aware of Japan's tactics of desperation as the war ground on to its finality. It also provided the U.S. with the inside information necessary to work out the mechanics of a successful peace treaty with Imperial Japan.

From then on, throughout the conflict, we knew what the Japanese were planning: what and where they were going to fiercely attack or stubbornly defend, and during the peace process, what the options were and which faction to support.

Less than six months after the disaster at Pearl Harbor, the battle of Midway took place; and there, by using the knowledge gained from our understanding of the Japanese code, the United States won that battle, our first win in the Pacific war. Then, by continuing to use the information obtained by studying the code, we went on to eventually win the war.

In spite of the good news, we had to keep our many successes gained through using the code absolutely secret. If the Japanese knew that we could read their code, they would have immediately changed it; and as we only found their naval code by accident, cracking a new code could have been extremely difficult.

As the war went on, and we were able to win many battles, one of Japan's seasoned diplomats, who had spent several years on duty at the Japanese embassy in Washington D.C., was said to have suggested to the radical officers' group, who were running Japan, that perhaps the United States was able to frustrate so many of Japan's military actions because they had broken the Japanese code. Fortunately, the fanatics running the Japanese war office scoffed at the idea that those "stupid Yanks" could have broken a Japanese code. And so, thanks to those fanatics, we were able to continue to use information gained from the Japanese naval code throughout the war in ways that eventually caused the defeat of those radical Japanese officers.

In addition to getting to know about acquiring the Japanese naval code, I learned about a most interesting foreign language and culture program that had been developed by the United States. From the Navy petty officers involved, I learned that, before the war, the Navy had begun an innovative immersion program to explore foreign languages and cultures, beginning with Japan. This program proved to be most helpful when hostilities began, particularly as it was a spectacularly successful supplement to the newly discovered Japanese code.

As it took place before the war, this program allowed a few selected Annapolis graduates to acquire a comprehension of the nuances of life in Japan by living there as if they were natives of that country. They were there not only to learn the language, but also to understand the culture of this exotic and secretive country. This cultural learning, together with their newly acquired language proficiency, helped in comprehending the complexities of the Japanese language, and in general, improving our translations of their communications.

Operative in rural Japan before the war, through a combination of efforts between the astute American ambassador to Japan, Joseph Grew, and the Navy Intelligence Office, this program allowed these special Annapolis graduates to live as if they were Japanese. They dressed Japanese-style, in robes and sandals, and resided in houses rented for them by Grew's staff. In these houses, selected non-English-speaking women cooked their meals and maintained the properties.

The American operatives also went to special schools selected by Grew, where they not only studied the Japanese language, but even the slang, as part of their comprehension of the culture of this exotic and secretive country. They became so proficient in the language that they often read the local Japanese newspapers to their maids, who were usually illiterate, as an education was not usually extended to lower class females.

This total immersion in Japanese culture, under the guidance of Ambassador Grew, was of tremendous value to the American Naval Intelligence Office by helping Americans gain a comprehension of the meanings implicit in the thousands of coded messages received, translated, and forwarded to Navy headquarters. This knowledge was of incredible importance in deciphering high-level military pronouncements; and, it made clear to the U.S. intelligence operatives the meanings behind announcements concerning the usually secret deliberations of the Japanese government.

At the time, all this restricted, and often hard to understand, knowledge was whirling around in my head, much of which I only understood later, as I was mostly involved in rigorous beach assault training maneuvers. However, it was obvious from my contacts with the Navy, and learning about these language proficiency and cultural awareness programs, that they were a most important factor in the success of the Occupation.

## Invasion Tactics

ON OAHU, WE TRAINED extensively for the coming invasion of Japan, including beach landings using invasion tactics, in addition to pillbox assault training. On special assignment, I often worked with the Navy on cooperative landing and assault training practices, and often heard background stories about what was really happening. Mostly, however, I trained with our Army units, including special amphibious training maneuvers, where I participated with my outfit, the Third Battalion of the 389th Infantry.

Leading up to one specific exercise, we were quartered overnight on troop transports. Sleeping shipboard in bunks did not provide restful sleep, and the noise from nearby crap games did not help. So, I was not feeling that great the next morning. We were awakened early and fed a breakfast of emergency rations that was endured, but not appreciated.

After breakfast, assembled by platoons, we were sent to the on-board staging area for landing maneuvers. This section of the troopship had cargo nets that had been hung over the side, with landing craft at the bottom of the cargo net. Like all the others, I was encumbered by my equipment: rifle, pack, helmet, cartridge belt, canteen, bayonet and entrenching shovel. I climbed down the cargo net, always a tricky maneuver. Sometimes, when the surf was high, I dangled out as the ship rolled in the ocean waves, and then sloshed back. Both the troopship and the landing craft rolled in the ocean waves; but unfortunately, in different directions, so that when you stepped off the net, you might end up dangling in the air. But, if the roll was in the other direction as you stepped off the cargo net, you might end up on the landing craft with a hard thump.

When our landing craft, a LCVP (landing craft vehicle and personnel), was full, our craft joined the other troop-filled landing craft, slowly circling, until all the landing craft for the battalion were loaded. While we were in the waiting circle, the landing craft were wallowing in the ocean waves at slow speed, rolling from side to side and up and down at the same time, causing a motion that was sure to induce seasickness. Once someone throws up, it is not long before others follow suit. This time, I did not; but it was not a happy time, even though this was just a maneuver and not the real thing.

When all of the landing craft in the battalion were loaded, we headed to the beach selected for landing. When the bow of the craft hit land, the front dropped down; and, no longer protected by the armor of the landing craft, we jumped out heading for the shore, even though this was just practice. We were splashing and wading through the surf, as the landing craft did not often put us ashore dry-shod.

After reaching the shore, the next step involved charging up to a pillbox, a partly concealed, mostly underground, fortified machine gun emplacement. In an actual battle, after cutting through the barbed wire, we would rush up to the pillbox while the enemy machine guns, theoretically, would be firing at us. In the midst of all this firing of guns, we would thrust an activated hand grenade or a pole charge into the aperture and, using the attached detonator (if it was a pole charge) set off the charge inside the pillbox.

While preparing for the invasion of Japan, we learned about the successful American invasive assault on the Nazi fortifications in Normandy, making way for the Allied invasion of Europe. In preparing for that invasion, ships from the American Navy blasted the Nazi fortifications on the Normandy coast with powerful salvos. When the Navy finished, the U.S. Army Air Corps followed up with numerous bombing runs. The effect of all this shelling and bombing? Not one of the Nazi fortifications was destroyed, and damage to them was minimal.

This left American foot soldiers with the task of attacking the Nazi machine gun-protected fortifications with just rifles and hand grenades. The main problem was that the Nazi fortifications were so protected by extensive barbed wire entanglements that getting through to the fortifications was just about impossible. The Nazi defenders' machine guns were trained on the entanglements; and if attempts were made to create a pathway through the entanglements using wire cutters, the soldiers attempting the cutting would be easy targets for the pillbox gunners. Unfortunately, all of the pre-invasion shelling and bombings not only had little effect on the fortifications, but the enemy barbed wire entanglements were also unharmed.

However, there was a little known, semi-secret device that was of utmost importance in clearing the way for the American riflemen: the "Bangalore Torpedo." This was a device created by the British Army in India, consisting of a long tube, originally bamboo, then steel, filled with explosives and armed with a detonator. Properly used, it was the best way to clear a path through barbed wire entanglements, as it did on the Normandy beachhead. An American soldier, lying prone for protection, would thrust the explosive tube under the barbed wire entanglement and then detonate it. A wide opening was then blasted in the entanglement, allowing the Americans to charge in with rifles and hand grenades, and take over the Nazi pillboxes. The casualties were high, as they always were when it was rifles against machine guns. But, with the barbed wire entanglements blasted away, the Americans could now reach the Nazi fortifications, and toss hand grenades through the apertures, blasting out the interiors of the pillboxes, leaving few survivors. With the Nazi fortresses knocked out, the Normandy Beach invasion, the largest marine assault ever created, opened the way for a successful invasion of what had been, up to then, Hitler's Europe.

As the Japanese shoreline defenses were known to be even more effective than those encountered in Normandy, our training for the invasion of Japan included the use of Bangalore Torpedoes, along with accelerated training in hand-to-hand combat and bayonet practice. But, left out of consideration was a potentially important weapon for combat in enclosed spaces, such as pillboxes and other types of fortifications that were in place to protect the Japanese homeland. This super weapon was the submachine gun.

## Lust and Leprosy in the South Pacific

BECAUSE OF THE HIGH rate of casualties in the Pacific campaign, particularly at Okinawa and Iwo Jima, the Army decided to bring in additional troops for the invasion of Japan. This extension of the war to the Japanese homeland was projected to be a real bloodbath. The Japanese army of 5 million, along with many millions of armed civilians, all of them ready to die for the Emperor, would, in the process, take not a few Americans with them.

Among the new arrivals for our battalion, Sgt. Martin Heffelfinger, a Pennsylvania Dutchman, had come up, we were told, from the Canal. This, of course, was not the Panama Canal, but the other canal, Guadalcanal. He had been stationed there, had seen combat in the area, and had contracted malaria. There was some question as to his fitness for additional combat, as he was still undergoing treatment for his medical problem.

When I joined the infantry at age 18, I had no illusions about it being an enjoyable experience. There was a war on, and you were expected to serve. While this rugged life does make boys into men - and it certainly influenced many aspects of my life - conditions in the infantry are always trying, and the rewards are few.

In the Army, the infantry, sometimes known as the "queen of battle," always takes the brunt of casualties. Infantry soldiers usually have to survive on tasteless rations; wallow around a lot in the muck; often have to sleep on the hard, cold ground; and march many miles encumbered with heavy packs, cumbersome rifles, heavy automatic rifles and back-straining mortar equipment. As it is definitely a hard life with few rewards, a somewhat apt saying of World War II encapsulated the general aura of that branch of service. It noted that if you were in the infantry, "you were fucked by the fickle finger of fate."

When Heffelfinger arrived, his longevity, along with his rating as staff sergeant, gave him seniority over all of the other non-commissioned officers; so I was concerned that I might be replaced as a rifle squad leader in my platoon. But Bruce Brusatti, the rifle squad leader of the Second Platoon, had a seniority rating slightly less than mine, and Heffelfinger replaced him instead. Our communications sergeant, Arthur Strauss, who was noted for his acerbic comments, summed up this new development: "Brusatti was," Strauss said, "fucked by the Heffelfinger of fate."

While sympathies were for Brusatti, Hefflefinger gained popularity; partly because he had been in combat and survived, but mostly because of his stories of life in the far southern islands of the Pacific. He was also a great spinner of yarns, usually in a light, bantering tone; and I joined the group that he had become the center of an offshoot of the

Post Exchange, the sergeants club, where he held court and told his stories. Heffelfinger had many stories he recounted for our interest, and we looked forward to hearing them, as the remote islands of the South Pacific were, in themselves, items of interest. One story that really got my attention concerned an officer, Lt. Herbert Johnson, who, like most of us, had been away from home for too long. He was stationed at Espiritu Santo, a remote French island that had an airbase important to the progress of the Pacific war.

Guarding the airfield from attacks by air or sea was an important component of the war effort. It was, however, a very boring existence for those, like Johnson, who found their routine patrols monotonous, particularly since all of the native villages, and their civilian populations, were "Off Limits to Military Personnel." There were at times some attractive young ladies to be seen, but always from a distance. While it had been rumored that some of the island gals had French bloodlines, and were said to be very enticing, not much more was known about them, as the no-fraternization policy restricted any acquaintance with these belles of the South Pacific. This was a situation that Johnson, like everyone else, experienced in our time in the Pacific war, as few female companions of any sort were available to the average soldiers or sailors ( or even officers) in the South Sea islands of the Pacific theater of operations.

Occasionally, however, young blood triumphed over regimentation. As narrated by Heffelfinger, this exception concerned the experience of the young lieutenant, Herbert Johnson. While driving by a native village when on a patrol, and very bored with celibate life in paradise, Johnson saw a hope for a possible evasion of the regulations that controlled his life. As he drove on by the village, an alluring young island beauty had discretely waved. Making it all the more provocative, she acknowledged his contingency, when he circled around, by looking directly at him with an appealing smile.

For the lieutenant, that smile had a 'come-hither' attraction that, after the long duration of his wartime celibate existence, was an invitation he could not resist. That evening, after his patrol returned to base, he commandeered a jeep and sought out the part-French exotic beauty. The meeting began well, and the initial mutual attraction quickly morphed into a passionate love affair. The lieutenant's passion for her began to consume him, and he made sure his continued involvement with his seductive young lover continued unchecked.

The lieutenant was able, with some stealth (and presents to the motor pool sergeant), to ensure that his frequent nightly appropriation of motor vehicles necessary for his transportation to his lover's native village went unreported. With after-hours transportation available, his visits became a series of clandestine lovemaking interludes, kept secret from headquarters.

Army rules and regulations insist that all rules always apply, even when there is no need for their application. Such was venereal disease inspection for officers and men having to be conducted on a regular basis, even on an island with a strict no-fraternization policy. For almost everyone on this particular base, the venereal disease inspection was a joke, as the prohibition against fraternization was strictly enforced( or so it was thought) by headquarters. The exception to the rules was the lieutenant involved with the exotic local charmer, as up to now he had been particularly adept at evading the regulations.

When it became the lieutenant's turn for the venereal disease inspection, it turned out that the joke, or rather the tragedy, was on him, as he had managed to have acquired a full-blown case of gonorrhea. As he stated this turn of events, Heffelfinger's tone of voice, usually bantering, took on an ominous edge. He went on to tell us that, as usual with cases of venereal disease, Johnson was asked to identify the person who caused the problem. At first, the lieutenant resisted, as he wished to shield his lover. But, when they asked him if this was a serious affair, he replied that he loved her, wanted to marry her, and eventually bring her home with him. When he was told that gonorrhea, if not treated, could make her sterile – but that it could be cured - Johnson relented, identified the young lady, and told them where she lived.

The medical department brought her in, confirmed her venereal disease, and was making plans for the start of treatment, when a general physical inspection revealed a further problem. To discuss this additional problem, the anxious young officer was called in to a conference room for a confidential conversation. There, he was told that, while she did have a venereal disease that could be cured, the disheartening news was that she also had leprosy.

*Leprosy*: as Heffelfinger said it, the word assaulted me. Just the utterance of the word, leprosy, and the way he said it, had the connotation of disfigurement and death. It was a malady without a treatment, except for isolation in a leper colony; and I was struck dumb by my primal fear of this dread disease. Leprosy was well known in Hawaii. On the nearby island of Molokai, a leper colony existed on an isolated strip of land, but it was understood that no one left that colony except in a body bag.

Before I could recover from the shock, and learn Johnson's fate, taps sounded, requiring my return to my barracks. While my mind was surging with questions about the fate of the young officer and his exotic lover, questions to Heffelfinger would have to wait until we could hear the rest of the story. I had hopes of learning more on the following day, but intensive training for the assault on the main islands of Imperial Japan began the next morning. This training, located away from base camp at the amphibious warfare center, was isolated quite far away from the main encampment. The isolation was necessary because live ammunition, often fired over our heads to replicate a real battle, made the area potentially dangerous.

After three days of this physically strenuous amphibious assault training, we were returned to our barracks at the main encampment. At my first opportunity, I headed over to the PX and to the corner there where the sergeants hung out, looking for Heffelfinger; but he was not there. I asked around, and finally found out disquieting news from one of the sergeants that his malaria symptoms had intensified, and he had been invalided back to a hospital in the United States.

Fine enough for Heffelfinger, but his absence left me with multiple questions to consider. Lt. Johnson's island lover would, of course, have to go off to a leper colony; but which one, was the question, as there were several. And would the young officer voluntarily go into exile with her, and what about his own condition? Leprosy, I had heard, is contagious from physical contact, and the young Lieutenant had had plenty of that; but it can take several years for symptoms to develop, if they do at all. In the meantime, would he have to live in a leper colony until his state of health was determined? Would this mean that he could never go home? I realized, as I stood there, pondering Heffelfinger's absence, that I would never know the answers to these questions about lust and leprosy. But I would always remember the frightening possibilities facing the young officer, and that this could happen to me, or anyone else. Isolated by war in a faraway place, danger of many kinds was always present. And now, far worse than the fear of death and dismemberment from the wounds of war, was the specter of leprosy.

## Friendly Fire

HEFFELFINGER'S DEPARTURE, AND THAT of others who had physical problems, was part of a greatly expanded preparation for the invasion of Japan. Also included in the preparation for the invasion was stepped up beach assault training, which usually came with over-our-heads firing of machineguns and trench mortars to create realistic training as we went through practice for the real thing. It was evident from this stepped-up training that the dreaded invasion was not far off. On one of these maneuvers, the training got out of whack, and we were subjected to an accidental death from one of our own weapons, an event classified as 'friendly fire.'

One of the saddest deaths from military action is that caused by your own artillery, or aerial bombs, or whatever. This is not a new problem. In World War I, the German infantry had a saying that summed it up: "We fear only God and our own artillery" - a sentiment, I am sure, that was echoed by all the troops in the death-dealing trench warfare of that horrendous war. In World War II, there were many examples of major losses from friendly fire. Some were reported, others were not, as deaths caused by friendly fire are something no military organization likes to admit, as it smacks of incompetence. However, in the realistic training for the invasion of Japan, most simulated battles involved the considerable use of live ammunition, sometimes with disastrous results. I was involved in one such friendly fire situation.

The use of live ammunition in training was not a new thing for the 98th Division: throughout our training, being fired on was par for the course. Early on, we had to crawl on our hands and knees for a considerable distance while machine guns fired over our heads. Machine gun bullets swishing through the underbrush near our in-training defensive positions was not a novel experience. Most of our training exercises where we acted out battle conditions involved our use of live ammunition against targets that simulated enemy soldiers.

Being run over by tanks was another way of getting us used to combat conditions. In the tank program, to introduce us to defensive tactics against tank attacks, we first had to dig a foxhole - nothing new there. Practically every time we were in a defensive position, we had to dig in; and an entrenching shovel was always carried in our pack as standard equipment. But, knowing that we were going to be subject to a tank offensive, where it was planned for the tanks to run over our foxholes while we were in them, stimulated us to dig deeply. When a tank rumbles over you, that is bad enough, but when a tank tread grinds directly over your head, that is a lesson learned.

In Oahu, where we trained for the coming invasion of Japan with beach landings and pillbox assault training, firing with live ammunition was often included in the training. This was in addition to extensive maritime assault training sessions, worked out in conjunction with the Navy, again with live ammunition. On special assignment, I often worked with the Navy on cooperative landing and pillbox assault practices; usually, however, I trained with our Army units. However, an unfortunate amphibious maneuver involving the Third Battalion incurred a friendly fire incident.

For this amphibious exercise, our squad, having landed and made its way past the beach fortifications, were told to wait in a designated area for supplies and reinforcements. As usual, it was "dig in and wait," which we were prepared to do. But this time, word went around that we should dig in quickly, and deeply, as live mortar shells would shortly be firing over us. This, they said, "was to give us an idea of the noise and confusion that went on in a battle."

In every maneuver, foxhole digging was mandatory. It was done in company with another soldier, using the buddy system: one soldier digging while the other guarded. In fact, Army life in general was practiced in the buddy system, particularly concerning sleeping conditions. Each soldier carried in his pack a "shelter half," that was half of a pup tent, along with one folding tent pole and six tent pegs. This meant that each G.I. had half a pup tent system and needed another soldier to completely install a working two-man tent. This fostered the buddy system in many ways, as it helped to like the guy you were sharing the tent with. As your buddy had the other half of a tent, when it came time to set up tents for the night, you and your tent mate would snap the tent halves together, set up a tent pole at each end of the tent, drive the tent pegs in, dig a ditch around the outside of the tent to prevent flooding, and crawl in for the night.

In preparing a daytime defensive position, one half of the two-man team would dig in while the other assumed prone position, "rifle ready" to guard against enemy attack, which was the way we were directed to do, all through training. According to the regulations, once the first hole was dug, the soldier safely dug in would be on guard, while his companion now dug his hole.

Our rifle squad was in an approved area, and we all fell to quickly, as the overhead mortar shelling had begun. We were either digging or guarding without much concern, as we had often gone through this overhead firing maneuver. This time, my partner, Arthur Rachlin, a good buddy, was in the last stages of finishing his foxhole, and I was stretched out, rifle ready, in the approved defensive position. So far, everything was working as planned.

In the location next to us, young Rudy Solis and his pal Kristen Barton were digging in, and on guard. They were a somewhat unusual combination. Solis, a happy-go-lucky Tex-Mex guy, was great to have around, as he always had a fun take on everything that happened. One of the new recruits, he was the assistant BAR man, working with Kristen Barton, the BAR gunner. Barton was a somewhat older, serious and stolid back-country man from Missouri. On the surface, Solis and Barton had little in common, but they were good friends and worked well together as the BAR team.

In every twelve-man rifle squad, the firepower was greatly enhanced by the BAR. Officially called the Browning Automatic Rifle, it was a combination rifle and machine gun. While lighter and more portable than the standard air-cooled machine gun used by the weapons platoon, the BAR was more cumbersome than the M1 rifle carried by the rest of the squad, with a heavier barrel and bipod legs. Because it was an automatic rifle, actually a hand-carried machine gun, it required considerable ammunition.

The importance of the BAR was that, as a portable automatic weapon, it could be carried wherever the rifle squad went, and would back up the squad's semi-automatic rifles with continuous automatic firing. Because of the extra weight, Barton had been selected to be the BAR man because he was sturdy and strong, and able to carry this portable machine gun for long distances. Like Solis, he carried with him additional ammunition, at his waist and draped in bandoliers over his shoulder.

Solis had almost finished his foxhole and was partly inside it, with Barton nearby in the guard position, when it happened: a mortar shell fell short, landing between Solis in his foxhole, and Barton, who was lying prone, but otherwise unprotected. The shell exploded as it landed, with shrapnel from the exploding shell spraying around them. Solis was hit just once and Barton twenty or thirty times, his body riddled with mortar fragments. Solis should have been safe, as he was quite far down in his fox hole and wearing a steel helmet. But one fragment from the mortar shell hit him in the head, just below where he was protected by the helmet. It was obvious that Solis was seriously wounded. Being close by, I yelled "MEDIC!" and rushed over to him. I quickly got a bandage from the medical kit we all carried and applied it to his wound. To no avail, he died there in my arms.

The next thing was to get help for Barton; and we did what we could for his many wounds as we waited for the medics. Barton seemed to be unconscious, and we discussed Solis' death, and whether badly wounded Barton was going to live. Finally, still in shock and with his eyes closed, Barton interrupted, choking out a few words. "I can hear you," he said, "stop talking about me." Fortunately, stretcher bearers arrived at this point and carried Barton off to the nearest road where an ambulance awaited. The injured and the dead taken care of, the maneuver was resumed, this time without the overhead firing of mortar shells, and without a great deal of enthusiasm. Barton's wounds were treated at the Army medical center, and he survived.

A few days later, at the cemetery at Schofield Barracks, a military funeral was held for young Solis, with our squad as part of the service. A Catholic chaplain said appropriate words, and then the squad, now seven men, lined up, raised their rifles, and at command, fired three volleys as a final salute to a fine young man. Rudolph Solis, a Texan of Mexican descent, was a good person and an excellent soldier, always cheerful, and always willing to do his share and more. It was a great loss to all of us, and I am sure, even more so to his family. Another good man, Kristen Barton, in spite of his many wounds, recovered and was invalided home. After the war, curious as to how the event had affected his life, I contacted him, and found that he had decided on a career in the ministry. Understandable, as his having been spared in spite of so many wounds, and Solis being killed while seemingly protected by being in the foxhole and the steel helmet and hit only once. This circumstance, Barton told me, led him to believe that he had been spared for a purpose.

The memory of Rudy Solis remained strong. After the memorial service, I wrote to his family about the high regard everyone had for him. And, on Memorial Day in 1946, I sent money to the Honolulu Red Cross for flowers for Solis' grave. I still have the letter the Red Cross sent me describing the floral arrangement they provided.

We never did hear what caused the short round. One rumor was that old ammunition that had been stored at Schofield Barracks was the culprit; but it just as well could have been a miscalculation by a mortar team who were not used to firing live ammunition over live soldiers. It is an unfortunate fact that mortar and artillery shells falling short, and other life-erasing incidents, are fairly common occurrences during wars and even peacetime maneuvers.

## Hotel Street, Honolulu

I DID NOT OFTEN GET from our base at Kailua to downtown Honolulu, but whenever I was there, the streets were crowded with khaki and white, as servicemen jammed the streets. Occasionally, a colorfully clad civilian, breasting the tide of uniforms, would dart by, seemingly in a hurry to be somewhere else; but that was not often. With many millions of men passing through on their way to Pacific battle zones, or stationed in Hawaii, Honolulu had become a key military outpost, a place where the local population was vastly outnumbered.

I did not see many young females among the few civilians that scurried around in the downtown area, mostly because there were not that many of them willing to brave the mobs of servicemen. Occasionally, they could be spotted among the sea of uniforms, but they were few and far between. This lack of available young ladies was the way things were, which soon became apparent to me, as I was not a commissioned officer - and without officer club privileges, I did not rate as a date.

Actually, most dating that happened in Honolulu included dinner and dancing in one or another of the mansions that had been converted to officers' clubs. It really didn't matter which club, as they all reeked with luxury and splendor - particularly the officer's club in the mansion that belonged to Doris Duke, which was the epitome of luxury. And, of course, as they were 'officers clubs,' you had to be a commissioned officer in order for you and your date to partake of the club's luxurious surroundings.

For the officers, their dates might well be service connected: nurses, Red Cross gals, or PX supervisors - on the whole, an upscale contingent of educated, smart, capable young women. Honolulu being an exciting place for someone to be at that time, they could experience a lifestyle that was a patriotic adventure. Because of their crucial work in hospitals and other military establishments, they were an important factor in the war effort.

There were also young gals who were local, and usually quite attractive; and an appearance by any of them in the military areas could cause a commotion, which usually kept them away from the downtown areas. Although of various ethnic and often Asian backgrounds, by virtue of the outstanding educational system of the Hawaiian islands, they were completely American in almost every way, although still clinging to vestiges of their ancestral customs. They quite often dated officers, as the palatial officers clubs, evolving from luxury estates, were an added attraction to be enjoyed as part of a dating milieu that existed for local gals and officers.

For the many million enlisted men on duty in Hawaii, or just passing through, there were bars, night clubs, and elemental post exchanges serving 3.2 beer, but not much action. Not having an upscale destination for dates kept the possibility of romance between servicemen and local gals at a low level. Occasionally however, there would be a romantic attraction between a G.I. stationed in Oahu, and a local female co-worker, but her family restrictions usually kept things from getting serious.

The celibate existence of the huge number of servicemen, most of them just passing through, and the possible problems that that existence could cause, led to the acceptance by the military command of a solution: brothels for the enlisted men. These were located in seedy hotels on Hotel Street, in the Chinatown area of Honolulu. These bordellos became well known to guys in the service, and were very popular, with long lines of servicemen waiting for their turn, the line extending up and down the street. This line was to be seen all day, but only during the day. As the gals worked practically non-stop, it was decided to close the brothels at night. This, so that the 200 or so strictly regulated, hardworking sex workers, could get some rest.

Hotel Street continued to be popular until late in the war, just before we moved to Oahu. So, all of what I have just discussed about Hotel Street is just hearsay on my part. In any event, Hotel Street's notoriety came to a reluctant end, when the wife of a visiting politician, observing the long line of waiting servicemen extending down Hotel Street, inquired into the cultural activity that drew so many avid followers. When the nature of the cultural activity was explained to her, she was shocked, and demanded that Hotel Street be declared "out of bounds for servicemen, in order to keep American boys, so far from home, from being exposed to depravity."

The diplomat's wife won her case. And with the brothels closed, the gals who worked there and the madams who ran the bordellos, financially secure from their years of hard work, retired to a quiet life back in the States. Or, if they so wished, they could continue to enjoy the comfortable Hawaiian climate by purchasing a villa in one of the Oahu suburbs, retiring to a life of luxury and financial freedom.

## The Atom Bomb and the End of the War

OUR ACCELERATED TRAINING FOR the invasion intensified our awareness of the high number of casualties from the various Pacific battles. This feeling was ever present, though seldom discussed. Particularly important to us were the latest casualties, those of the Okinawa campaign. The highest numbers since Guadalcanal, they gave us a foretaste of what was to come.

Even after their loss of Okinawa, the Japanese military dictatorship continued to maintain their fixation on their belief that the only honorable deaths were in battle or by ritual suicide, and this belief would never allow considering a surrender of any kind. The American war cabinet intensely studied the no-surrender attitude of the Japanese military, which we were well aware of through our ability to decipher the Japanese code. This was combined with the realization that making an assault on the Japanese homeland would not only cause enormous American military casualties, but also millions of Japanese civilian deaths and injuries, the practically complete destruction of the cities of Japan, as well as the annihilation of the many-million-strong armed forces of Imperial Japan.

But, when it came, the incredible news of the atomic bombs and the subsequent surrender by Japan caught us by surprise. Our training for the landings and assaults on Japan's formidable defenses had been intensive, and the invasion of Japan's home islands would probably have caused extensive American casualties, as our recent losses on Iwo Jima and Okinawa had shown. The inner feelings of apprehension that dogged us, concerning our probable casualties in the invasion, were seldom discussed, but had always been there: the uninvited elephant in the room.

But now, the world had changed. With the war over, we might be going home after all, as it was to be an "occupation" and not an "invasion." It was a head-spinning reversal of the probability of our negative chances of surviving the tenacious Japanese resistance to an invasion of their homeland.

Immediately, preparation for the invasion ended and groundwork for the Occupation began. In a short time, our modus operandi changed, and we were on our way to a disparate focus in a different world. The war materiel we had just loaded was unloaded from our ships, which then were reloaded with occupational supplies, as well as a substantial amount of provisions, because food supplies for Japanese civilians were reported to be at a low ebb.

All of the changeover of cargos took some time, and as the tensions and forebodings evaporated, we could relax somewhat. Discussions began about the end of the war, and we began to hear scuttlebutt about what had really happened. And, as nobody seemed to know much of anything about the nuclear bombs, questions kept coming up about them, and what part they played in the surprise ending of the war.

The answer to questions about the end of the war was that, while it was a long and complicated story, it was well worth the effort to explore what it was all about. Questions about the nuclear bombs went mostly unanswered, as many of the details were off limits at that time. But, I continued to be selected for special projects and eyes-only episodes, as well as involvement in several situations that I only learned what they were all about later on. I gradually became aware of what all the end of the war maneuvering had been about. Basically, using the bomb seemed to be a way to stop the killing by ending the war; and, as all of us foot soldiers would have been cannon fodder in the invasion, what went on at the high levels of command was of supreme importance to us.

Before the war was ended, of grave concern was our high casualty rate from the various Pacific battles, particularly since the Japanese would be defending their homeland. The battle at Okinawa was a perfect example. It was considered one of the most intense of WWII with Army and Marine ground troops losing 49,151 killed or wounded. Mostly because of the Kamikaze attacks, the Navy casualties were also significant with 34 ships sunk, 368 damaged, and with more than 4,900 sailors killed or missing.

The casualties at Okinawa for Japan were also high: very few Japanese soldiers survived the battle, and over a third of the island's civilian population were killed. The deaths for Japanese soldiers killed in battle were officially counted as 107,539, while hundreds of Japanese soldiers and civilians followed the example of the losers of the battle on Saipan by committing suicide in caves or by jumping off cliffs into the sea.

However, intense study of the Japanese codes revealed that, even after the loss of Okinawa, the Japanese military dictatorship continued their fixation on a belief that the only honorable deaths were in battle or in ritual suicide.

There was a non-combat, but still a military, factor that weighed heavily on the decision to attempt to end the war by dropping the bomb, before attempting the actual invasion of Japan. This was a directive, found in captured Japanese military archives in the Philippines, that applied to all Japanese commands. This directive mandated that, immediately upon the initial physical American landing on the Japanese homeland, all prisoners of war were to be executed. This included 50,000 American prisoners along with a greater number of English and Dutch and Asiatic prisoners, for a total of 140,000 prisoners of war. In fact, it was rumored that some executions had already commenced.

There were many theoretical reasons for ending the war through the use of the nuclear bombs: mainly that the use of this weapon would convince the radical Japanese generals that they should commence peace negations. However, the first atom bomb had little effect on their thinking; and the destruction of Hiroshima did not seem to matter to them or convince them that they should even consider negotiations. We knew from intercepted radio broadcasts that they were adamant believers in the no surrender fixation, even if the entire country was on the verge of total destruction.

The radical generals who were running Japan did have a solution born out of desperation, which was to request the Soviet Union to join them in resisting the American invasion. This was pure fantasy on their part, as the Soviet Union was already moving toward a complete takeover of Japanese Manchuria - which they did, shortly after the first nuclear bomb was dropped.

The antipathy of the radical generals toward any consideration of ending the war, along with their maneuvers toward a Soviet partnership, were well known to the U.S. through the deciphered code messages. This attempt to ally Japan with the Soviet Union proved the intransigence of the Japanese generals and the necessity for the second bomb.

However, even the second nuclear bomb was not enough to sway the radical military. The success of the American military forces, on land and at sea, and the subsequent destruction of so many Japanese cities, made the importance of ending the war obvious to many Japanese. Our military intelligence experts had known, from our reading the Japanese codes that the Emperor cared about the people of Japan and was in favor of ending the war, never having been in favor of the continued military conquests.

During the air attacks when Tokyo was fire-bombed, causing tremendous destruction and 100,000 casualties, the area around the Emperor's palace was exempt from attack by the United States. We wanted to make sure that the Emperor survived, because the American government knew that we needed the Emperor's help in convincing the Japanese to end the war.

The big breakthrough came from the Potsdam Declaration. The Potsdam Declaration, by America and our allies, began to have an effect within Japan. It actually was aimed at the Emperor, and was accepted by him, along with the unconditional surrender requirements of the Declaration. The Emperor used his power as the spiritual leader of Japan to activate the ending of the war. He did this through a tape recording which was to be broadcast the next day, announcing his decision to abide with the demands contained in the Potsdam Declaration.

When the radical militarists, who were violently against surrender, found out about the tape, they went against every precept of the Japanese culture by attempting to negate the Emperor's projected declaration of the end of the war. They planned to do this by finding and destroying the tape, and thereby forcing a continuation of the war. In the process of the search, various groups of radicals took over the palace grounds, even the palace itself. However, the tape, reportedly hidden in the Emperor's wife's bathroom, was not discovered. When it was broadcast as planned, the day after it was made, many of the radicals, with their passion to continue the war thwarted, followed the dictates of their Samurai code and committed ritual suicide.

The Emperor's speech was broadcast to the Japanese nation, with the Emperor stating that, because of his concern for the Japanese people, he was declaring that the war was over. Because the God-like Emperor had said the war was over, the end of the war was accepted by the vast majority of the Japanese people. And. because the announcement came directly from the Emperor himself, the people of Japan worked in harmony with the Occupation and treated the occupiers in a friendly manner.

This cooperation was of tremendous importance to all of us who would be involved in this new concept. Probably for the first time in history, Occupational forces assisted in the recovery of a defeated enemy. Instead of demanding reparations, the United States helped Japan recover from a disastrous war. The astounding results of this reversal of fortune for Japan, changing it from a brutal dictatorship to a peaceful and prosperous democratic entity, were so remarkable that a similar set-up was created for Europe, there called the Marshall Plan, where it achieved similar results.

# Part Five:
# Occupation of Japan

## On Our Way to Japan

IT DID NOT TAKE long to adjust our thinking to our new way of life once the war was over. Up to this point, there had been the omnipresent specter of war, with death and dying imminent possibilities. But the post-war existence was so radically different, it was, for us, a new world to experience. When we were preparing for the invasion, every aspect of our thought process was spent in preparing for war; and now, none of it was. Instead of preparing to invade Japan, we were on our way to occupational duty in the homeland of our former enemy. All of our training on invasion tactics (including landing while under fire, which would have caused a shocking number of casualties to happen in a real battle) was no longer necessary. Because now, it was a whole new ball game, where the landing might prove to be exciting and enjoyable, rather than grim and dangerous. Apprehension had been replaced with anticipation; Tokyo Rose had been replaced with Madame Butterfly.

The 98th Division's part in the invasion was to have been implemented by using the divisional troop carrier concept, a military strategy that kept the troops and equipment of a division together, to increase the firepower of the group. Our troopships, three APA's (army personnel carriers) and, for artillery and supplies, an AKA (army cargo carrier), comprised our small fleet. In addition, the troop ships carried on their decks landing craft ready to be used for beach landings. Together, this comprised the 98th Division invasion force. It was a way to keep the Division's firepower working in unison during landing. This new concept of coordinated landings had been created for the invasion of Japan, and the 98th landing force was one of many similar landing forces that were being assembled for the invasion. But then, the war ended.

As our troop ships and auxiliary transports headed off for the Occupation, I began to need some new ways of thinking, mostly about what might have been, and what was going on now. If our voyage had been part of an actual invasion, we would have been traveling under blackout conditions and in zig-zag formations, as we would have been in constant danger from submarine and air attacks. Along with that, we would have been in a kind of blackout of personal feelings, because of apprehension about our expected plunge into the presumed carnage of the invasion. But now, with apprehension fading away, our voyage became a peaceful, if not very pleasant, oceanic excursion.

The same 'blah' food and restricted menu in place in war-time voyages were still in place; but somehow, it was not so bad, now that blackout conditions no longer prevailed. However, the extreme heat discouraged anyone from lingering in the mess quarters. Another negative was the way food was served. After standing in line for some time, you were given a tray and a dish, and you were then served by having the dish on your tray filled with whatever the specialty was for that meal: everything dumped into that one dish, including the dessert - usually canned fruit. You then went over to the start of one of the long tables where, because there was no place to sit, you would stand as you ate. Then, as new arrivals moved on to the table behind you, it would be necessary for you to slide your tray along, eventually all of the way to the end of the table - with all of this going on as you ate, still standing. You needed to be a fast eater, as otherwise, you would have been pushed off at the end of the table, whether or not you had finished eating. The push would be by other G. I.s, whose trays were being pushed along by new arrivals moving in behind them. So, you learned to eat fast. The meals were usually okay; but as only two meals a day were provided, and with no opportunity for snacks, you made sure that you cleaned your plate before you reached the end of the table.

Every day aboard our troopship, the *USS Lattimer,* was not much different from every other day. The sea was quite calm, the weather, hot but bearable. There was nothing much to do but wait, play cards, shoot dice, and read. I tried playing a few games of checkers, but the country boys, who were obviously skilled at checkers, beat me to a frazzle every time I played with them; and as I didn't enjoy being consistently humiliated, I gave up on checkers.

Below deck, where the sleeping quarters were, was usually crowded. Canvas bunks were stacked in rows that left little room on either side, or between you and the man in the bunk above you or below you. There was not a lot of breathability below decks, as the air there was usually thick with cigarette smoke, and crowded with dice and poker enthusiasts, who always seemed to be smoking. The American government made sure that cigarettes were always available; and the standard ten-minute breaks were announced every hour over a loudspeaker, telling us that "the smoking lamp is lit."

Fortunately, we were allowed to sleep outside on the deck, which I often did by spreading my blanket under one of the landing craft that were stacked on the deck. These were stacked in layers, as they were always ready for action, and could be launched right from the deck. The hard deck under the landing craft was preferable to sweltering in the sleeping area below, breathing the fetid air and smoke. Fortunately, with the war over, discipline was now somewhat relaxed; and in the high heat, shirts sometimes could be dispensed with.

Now that we were off to a peaceful entry to an exotic Japan, and with time on my hands, conversations with my sea-going grandfather came back to me, and I remembered hearing from him about incidents that involved colorful circumstances. Because I would have been part of the invasion, I did not expect to see or hear anything of old Japan. But now, at least part of what he had weathered might still be there; and I was looking forward to experiencing his version of the mysterious Orient.

Japan was about 1,485 miles away from Oahu, and the scuttlebutt had Kobe as being our eventual destination. The sea was almost mirror-calm; and it was very hot - not just the air around us, but the sky itself looked hot. I was reminded of the verse in Coleridge's *Ancient Mariner*:

> *"All in a hot and copper sky, the bloody Sun at noon, right above the mast did stand, no bigger than the Moon."*

For some reason, we did not go directly to Japan; Saipan, we were told, would be our next destination. A former German possession, this fertile island had been allocated to Japan at the end of WWI as the reward for their declaration of war by Japan against Germany. Since then, Saipan had become a fortress island, bristling with gun emplacements and fertile plantations that were the source of much of Japan's sugar. It was the site of a recent, very fierce battle, where the United States Marines gained a victory, but not without considerable casualties. The word was that it was to be our only stop along the way, but there was no information on why we were stopping there. As Saipan had been a heavily fortified Japanese stronghold, the debris of war were everywhere. Where we came ashore, we passed several wrecked Japanese landing barges, rather battered, to be sure.

With some pals, I stopped off at a navy recreation joint where we were served some American beer, which was enjoyed, even though it was the usual post exchange 3.2% low-alcohol variety. Now, fortified against the heat, we walked through what was left of the town. Some of the remains of the garrison's buildings, roofless and crumbling, showed the intensity of the fighting. The sugar mill tottered on its steel frame, with portions of the roof and one wall remaining. At the water's edge, on nearby reefs, knocked-out American amphibious tanks gave evidence of the violence of the battle. Along the shore, rows of Japanese pillboxes remained, although in dilapidated condition.

And then, as we walked around, we saw much evidence of the battle, including the large and well-tended cemetery of the 4th Marine Division. The main town was clobbered and deserted, as the Japanese civilians who survived (and there were not many) were now confined to an enclosure of some sort. As the battle had raged on, many Japanese soldiers who, in spite of their stubborn and fanatical resistance, eventually realized that Saipan was lost. They then exercised their no-surrender mandate, with large numbers of them, and many civilians, committing suicide by jumping off a high cliff into the sea.

This action brought to mind the thought that, if this was the reaction to losing on this remote island, isolated by many miles from the home islands, what could we expect the reaction would be to the actual surrender that had just taken place in their homeland? This was a question that no one seemed to have an answer for. But, official concern was shown by our being told that when we landed in Japan, we would be combat-equipped, ready for any resistance. After digesting this preview of the possible reaction to our imminent occupation of Imperial Japan (an empire that had never been successfully invaded), we re-embarked, and were now on our way to the long-awaited destination: Imperial Japan.

The last night before reaching Japan, the weather, dark and uncertain with the possibility of a severe storm ahead, mirrored my feelings about the immediate future. The way ahead was uncertain, as we had no way of knowing what our reception would be, particularly from the radical officers group who had initially engineered the takeover of the Japanese government.

As our ship plowed through the murk of the gathering storm, we seemed to make little headway against the total darkness of the inseparable sea and sky, as the dark night pressed in on all sides. Now and then, one of the ships' searchlights, perhaps probing for mines, flitted nervously ahead, as though echoing the restive feelings of those on board. In the course of its scanning, the searchlight's beam hit a glancing blow on something, bounced off, then caught again, and held in its beam a small Japanese fishing boat, a motorized sampan, drifting aimlessly with the current. As we came closer, the crew could be seen, some lying on the deck, hardly moving at all, some at the railing, waving listlessly. The small boat was obviously adrift, without power, and probably without sustenance. We guessed that they didn't know that the war was over, and that perhaps they thought we may try to sink them. But, in any event, they must have been desperately in need of rescue.

Our searchlight followed the sampan as it drifted passed us, rocking uncertainly in our wake; and then, as our ship sailed on, the small boat faded away into the outer edges of the searchlight. Our searchlight snapped off, and then came back on, catching the sampan again, the crew now waving frantically, and crying out in desperate voices as the sampan faded further astern. Then, as we sailed on without slowing down, the crew of the sampan stopped calling, standing mute and motionless until we were too far away to see them distinctly. By then, the searchlight had moved away from the small boat, and had returned to its nervous probing of the sea ahead.

With a major storm about to descend on us, we knew there was no way our troopship could have interrupted its rendezvous with the rest of our convoy to rescue the crew of the sampan. By this time we were all standing on the deck, looking at this unfinished drama, wondering what would happen next; and the possibility of rescue was discussed among us. We hoped that it would happen soon, particularly since the gathering storm was closing in. But there was hope, now that the position of the disabled sampan was known, and probably reported.

As the storm gained strength, it became impossible to stay on deck, the force of the storm sending us to security below. In the morning, however, through the downpour, we could dimly see the ominous coast of Japan looming. It was not Kobe, as had been rumored, but Wakayama, we were told, the port city for Osaka, the second largest city in Japan, and our destination.

Even though the storm had freshened into a typhoon, with howling gales and drenching rain, the landing plan was still to be carried out. Wakayama, like most of the ports in Japan, was mined, so we made a beach landing in spite of the storm. Dripping wet, we climbed down the cargo nets to landing craft that had been perched on the deck of our troopship and were now waiting to take us ashore. The loaded landing craft then lurched toward land, through the surf, depositing us on to a strange land and a new life as soldiers of the Occupation.

## Landing in Imperial Japan

IN SPITE OF THE howling gale and drenching rain, the landing had to be carried out, as it was part of a total plan for the Occupation. So, wearing steel helmets and backpacks, with rifles strapped across our backs in order to have our hands free, and already soaked from the storm, we clambered down the slippery cargo nets into landing craft. While combat-style landings in rough weather were always somewhat dangerous, at least with the war over, no one was shooting at us. As the landing craft brought us close to land, the front section opened and we jumped out, wading ashore through heavy waves. Once ashore, we were formed into rifle squads, preparing for any event. We were the first Americans to land in the area, and not knowing what the reception would be, we were ordered to fix bayonets, and proceed through the village as if on a combat patrol.

By squads, we then fanned out through the totally dark village. The small seaside houses were not much more than shacks, without real windows - just shutters, now tightly closed, with no glimpse of light from the inside, or any sound of human activity. The rain had let up, but it was a moisture-laden, murky day; and while it was quiet, it was, in fact, super-quiet. As my squad moved out along a narrow lane, there was a feeling of apprehension in the air. This coincided with another strange impression that gave me an eerie feeling: because it was so quiet, it was as if no one in the area was still alive. My guess was that the people knew about the havoc that ensued when their soldiers stormed ashore, and they probably expected the same from us, and so stayed out of sight and sound.

As my squad moved out in the approved single file, combat formation, with other squads filing down other lanes, the unnerving death-like stillness surrounded us, until it was unexpectedly broken by the clatter of sound against the pavement. I was standing behind a building at an intersection waiting for the scouts who had gone ahead to give the okay to cross, when I heard the clattering sound. As I looked out from behind the building for the source of the noise, I saw that it was a small boy running down a side street, in my direction. As there was no other person visible, he probably never should have been out, and was looking frantically behind him as he ran in my direction, obviously in an effort to evade another of our patrols. Just as he reached the corner, I stepped out from behind the building that had been sheltering me. Confronted by an armed soldier, an obvious enemy, he ground to a halt; probably expecting instant death, his face took on a sort of sickly green. Trembling and shivering, he tried to stand at attention, and gave me a shaky salute. As I returned his salute, he somewhat recovered his composure and was off with the speed of light. But he was not gone forever; we would see him again the next day.

After our patrol of the seaside village, we bivouacked in a small grove of spindly trees; and while the rain had stopped, everything was damp and drizzly, including the blankets from our packs. Our first night in occupied Japan was a pretty miserable experience, as we were completely soaked and, of course, concerned about what would happen next.

We later learned that the beach houses were part of the city of Wakayama, which, before the war, had a population of about 200,000 people. Now, we learned, Wakayama was no more, as most of the old city, consisting of two-story wooden buildings, had been destroyed by fire bombs. The exceptions were an imposing, European-style bank that was still in operation, and a few other fireproof office buildings. On the beach side of town, some of the older buildings still survived, but most of the inner city had been burnt out.

The rain continued off and on, most of the first day ashore; but when it stopped raining, the air was crisp and clear, and the scenery: gnarled pine trees straggling amidst sand dunes, and picturesque ocean-side beaches, reminded me of Cape Cod. On the inland side of the dunes, rice paddies stretched on to where they nestled up against the side of a low hill, with rugged mountains towering above the hill and the cultivated fields. On that side of the dunes, away from the ocean, the air was not so good, as the rice paddies, like all Japanese farm areas, were drenched with human waste.

In the morning, while we were opening and eating our rations, who should appear but my audacious young friend from last evening's patrol. He kind of sidled in, looking longingly at our rations; but when he was given a small donation of food, he did not gulp it down. Instead, he tucked it into his shirt and quickly disappeared, probably back to his home.

He must have assured the villagers that we were not monsters, as in a short while, a hesitant delegation of elderly stragglers approached in a somewhat fearful, yet placating way. They bowed to us, and said some things in an anxious way, so it seemed. Even though we had no idea what they were saying, we gave them some food and a few cigarettes, which we always had a plentiful supply of, as they were packed in all of our rations. Obviously delighted with these small treats, the elders went off, clutching their prizes. We, not knowing what a Pandora's Box we had opened, thought we had seen the end of visitations by the local people.

Americans had been vilified by Japanese propaganda during the war, and the Japanese people had been taught to fear us. And, while the wholesale atrocities committed by the Japanese military in China, and wherever the Japanese had taken over, were seldom discussed in Japan, most people were aware of unpleasant incidents that had happened in conquered lands, and might happen to them now in their own country.

However, the American policy of strict discipline, rigidly enforced by General MacArthur, prohibited any misuse of military force, or any unlawful activity. This ensured a crime-free American Occupation, which, in combination with the Emperor's proclamation ending the war, made the Occupation a relatively problem-free situation for both the occupiers and the occupied.

Food in Japan was extremely scarce, and hunger evidently had overcome the locals' fear of these strange invaders. In a short while, many more people came to visit us, singly and in family groups. As long as we camped there on the beach at Wakayama, they were there when we bivouacked for the night, and when we got up. They watched us while we worked assembling occupational equipment - and particularly, they watched us when we ate.

The Japanese did not seem to have benches or chairs - they just squatted as they watched while we worked, just as they squatted while they were waiting for buses, and other of the very scarce sources of transportation, or on any occasion where they had to wait. They watched us whenever we opened a can, and when we took a bite of rations, they would suck in their breath, halfway between a hiss and a groan.

Animals in a zoo could not have aroused greater curiosity than the American military did for the people of Japan, particularly when eating. They soon made themselves at home with us and our equipment, fearlessly examining our tents and workstations, and even invading our hastily dug latrine, the Japanese having had no modesty about such things.

Villagers kept appearing, until it got to the point where we were being overwhelmed by the mass of people surrounding us. Eventually, we were able to get a message to the village elders saying that the people should stay away. A Japanese police official was posted to ensure that we were not bothered, and that worked for a while. But late in the afternoon of the next day, our presence, and the knowledge of available food, was too much of a lure for the kid who had been our first friend in Wakayama.

In spite of the orders to the villagers to stay away, he tried to sneak into our bivouac again. He did not get away with it, as he was quickly spotted by our guarding cop, who called to the boy, evidently summoning him to approach. The boy marched up to the cop and stood at attention while the cop was berating him, which we could tell from the cop's tone of voice. Then, after berating the boy, who still stood at attention, the cop smacked the kid across the face, knocking him down. The boy stood right up and was slapped down again. At this point, we began to loudly complain; and the boy and his captor marched away. We never did know what happened to our small friend, as the next day we moved on to our destination, Osaka.

We later learned that this treatment of the boy was an example of the strict police discipline common in Japan, where few human rights were observed. The swords that the police carried were sometimes used to speed up laggard pedestrians - not by stabbing the laggards, but by slapping them with the flat of the sword.

Before long, the local people returned to a vantage point in an area just outside of our bivouac; and while they refrained from close contact, they continued to observe our activities. It was apparent that their attitude was one of curiosity rather than anger or fear. And, we eventually realized that this attitude on the part of most of the civil population was due partly to the discipline of the Occupation; but most of all, due to the radio broadcast by the Emperor, ending the war. When he said the war was over, it was over.

## Amphibious Vehicles

AFTER THE INITIAL LANDING of our troops at Wakayama, all supplies to sustain the occupying forces, and to relieve the imminent starvation in Japan, had to be brought in from cargo-carrying ships. As the harbor at Wakayama was extensively mined, with no working wharves available, everything had to be brought to shore by small craft that ferried the supplies in from ships that were anchored away from the wharf area. Once landed, the supplies would then have to be brought to some depots that had been set up in Osaka for military equipment, and to others that stored food for the civil population.

The military solution for the problem of unavailable wharf facilities was an outstanding American invention: amphibious vehicles. These 2 ½-ton trucks, (officially the DUKW, but known as Ducks), were an American specialty (along with amphibious tanks and jeeps). These seagoing trucks were unusual vehicles that could maneuver at sea as well as on land, but had never been seen or even imagined by the Japanese people. In fact, all information about America had been severely censored during the war, and so, much of our equipment was new to them. Just as everything Japanese was completely strange and exotic to us, in every way, we were completely strange and exotically foreign to the Japanese.

The local population, there on the waterfront, curious about us and what we were doing, kept crowding in on us, examining every activity. However, their attention was diverted when the first of a long line of small boats was seen heading from our cargo ships through rough water, toward the nearby beach at Wakayama - the place where we had first landed.

There was nothing new about cargos being transported from ships at sea to the shore by small boats. But when these boats came up out of the roiling surf and continued on land as trucks (proving to be seagoing trucks), murmurs of astonishment rose from the Japanese civilians gathered nearby. They were very curious about this strange new sea-and-land military vehicle; but even more strange to them was the outlandish sight of the drivers of these amphibious trucks: Black men. Seagoing trucks were astounding enough; but when these amphibious vehicles were driven by Black men, the combination must have been, for the Japanese, almost like a scene from outer space.

Japan had always denied any person not of the Japanese race to be allowed to live in the country, except for a few Koreans. Dating from the Japanese occupation of Korea in 1910, some Koreans were allowed into Japan to perform menial tasks, similar to the functions of the Untouchables in India. No other foreigners were allowed into Japan, except as tourists; and so, an outsider, particularly a Black outsider driving a seagoing truck, was for them a scene almost beyond belief.

Because of segregation policies in the United States military, instituted in WWI by then President, Woodrow Wilson, some Black soldiers were allowed to serve in the American Army in both wars, but they were not permitted to carry arms. They could, however, be laborers, mechanics, or truck drivers. However, in WWII, they were permitted to be drivers of a new vehicle, the amphibious truck. The Black soldiers, although limited in their duties, served with distinction. The only deviation from the military segregation of Blacks by the American military in WWII was created for publicity purposes. A small group of Blacks from Tuskegee Institute, a well-known Black college, were allowed to become officer/pilots, but only as a one-time, separate unit in the Army Air Corps.

In general, the Occupation, so foreign to their way of life, must have been a polarizing experience for the Japanese people. The total American Occupation, with its technical innovations and democratic idealism, was, in itself, unnerving and astounding to them, as Japan was a country that never, in its long history, had had any civil rights, or even any consideration of them. The Japanese were not only unaware of what was happening in the rest of the world, including the defeat of their Axis partner, Nazi Germany, but the populace had been kept ignorant of Japan's many losses in the war. And so, the unconditional surrender by a nation of people who had never surrendered was an incredible shock to the psyche of the people of Japan.

It was also pretty astounding, and memorable for me, as an American soldier who had some background in this bizarre country, to observe what was happening. The tableau of this scene from war-torn Japan resonates in my mind's eye as if it happened yesterday. I will always remember the anxious spectators in their colorful garments, the somber ocean, with the ships at sea unloading their cargos into small craft, which headed toward the surf-tossed beach where we had landed several days ago. And then, the small craft wending their way through the surf, on to land, then going on past us, now as seagoing trucks as they continued on land, past the bewildered Japanese. The amphibious trucks on their way inland to their destinations, military and food depots, produced a dazed response from the civilian onlookers.

And then, there was my own electrifying reaction to having been involved in this unforgettable scene. But, however unusual it seemed to be, it was just a sample of the extraordinary events that I would witness during this unusual occupation of this land of bizarre customs and strange occurrences.

## On to Osaka

AFTER COMPLETING OUR ASSIGNMENTS at Wakayama, we then moved on to Osaka, the second largest city in Japan, and the headquarters for the 98th Division's portion of the Occupation. As soon as we arrived in Osaka, we were immediately assigned objectives to attend to, including the destruction of enemy weapons and many other varieties of war materiel. After our clean sweep, which took some time, nothing of any Japanese military importance was left.

When we moved to Osaka, our destination was former military barracks just outside of town, which, while clean, were quite primitive. Japanese soldiers evidently were provided few material comforts. The barracks were without heat, and as Osaka's latitude was about that of Washington D.C., it could be cold and rainy during the winter months. But the lack of any creature comforts, as demonstrated by the primitive living conditions in the barracks, showed that the Japanese military evidently had a simple, pared-down existence.

In addition to the lack of heat, the barracks also were without running water or indoor toilets. There were buckets provided for the collection of solid waste for distribution by ox cart, to local farms as fertilizer. Open air urine cesspools relied on evaporation to keep from overflowing. But the odors from hundreds of years of urine were powerful. After we took over the barracks, kerosene stoves were installed to provide heat, water was piped in, and the toilet problem was remedied by the installation of the American army's portable toilet system, making things reasonably comfortable, and somewhat odor free.

Ever since arriving at the former Japanese military barracks, we spent most of our time cleaning, painting and renovating the elderly quarters, with very little time off to explore the surroundings. However, even though much of Osaka was in ruins, when I eventually ventured downtown, an informal kind of commerce was obviously thriving. Amid the debris in the downtown area, wherever there was room for a stand or a booth, or any available place at all, merchandise (black market or otherwise) was shown, and various foodstuffs were available.

Venturing downtown whenever off duty, I wandered through the narrow streets looking for possible art treasures, fascinated by the constant activity and colorful street scenes. But there was not much of interest to acquire, except for occasionally, a few samples of textile art. A major exception came from a dingy shop, where I spotted, off in a corner, a model of a Japanese battleship with meticulous detailing. It was expensive, but as my father collected naval memorabilia, I felt it a worthwhile purchase.

After a few weeks of intensive Occupational duties, a deviation from work was announced: a lecture on the history of Japan, to be given at a local university by a noted professor. As I was interested in the subject, and transportation was provided, I decided to attend. Transportation, army style, was, as always, a large truck in which you stood, as seats were never provided. But at least it was free transportation. Several of us went, and as it was a tiring ride, I was grateful that seats were provided when we arrived at the lecture hall.

Japanese history was a subject that I had considerable interest in, through family connections, but the lecture, translated for us, offered a very strange version of history that did not come close to what had really happened. According to the professor, Japanese people were never warlike, and Japan was always a land of peace and harmony. He went on about peaceful Japan for some time, which may have been true when the Shogun ruled Japan, several hundred years ago. It was, then, a hermit kingdom where no one could enter or leave, and the samurai ruled with an iron hand under the overall authority of the Shogun. In the professor's version of peaceful Japan history, this had been a country with no conquests of foreign places.

However, while the professor's image of peace and quiet may have existed during the distant past, the takeover of Japan by radical military elements had changed everything. Once the Shogun was ousted, this new direction led to the conquest by Japan of much of the Far East, all of which began in fairly recent times. These new military conquests included the takeover of much of China, Korea, Formosa, and many of the islands of the Pacific. But, none of this military activity was mentioned in the lecture; it was as if the modern conquests had never happened. The professor painted a pretty picture of a long-ago time and place that had for many years been pushed aside by the radical elements.

The lack of any factual information made the lecture a waste of an evening, but to make things worse, much worse, on the way back to the barracks, a red glow in the sky could be seen. The glow got stronger as we neared our barracks; and in fact, the red glow was from our barracks, which, by the time we got back, was blazing fiercely. One of the kerosene stoves that the U.S. Army had installed had somehow tipped over, and the aged, extra-dry wood of the barracks quickly burst into flames, causing our home away from home to be totally destroyed in a very short time. Everything I owned went up in flames, including all of my army gear and all of the art treasures I had collected.

It took a while to get things straightened out, and military gear replaced, but all of the personal possessions were gone for good. Everything was lost as the paint-soaked dry old wood of the barracks had gone up like a torch. We were given new clothing, but nothing could make up for all of the personal stuff I had accumulated since coming to Occupied Japan. All my photographs were gone, and that was bad enough; but all of my paintings and sketches I had made while in Japan were also destroyed.

However, in spite of the fire, Occupation duties continued, and in between these duties, I did some sightseeing. Although the main business section of Osaka had been pulverized by the bombing, some modern fire-proof buildings were still standing. As in most Japanese cities, downtown had consisted of many two- or three-story wooden buildings, most of which had gone up in flames. Amid the debris, the still standing fireproof buildings stuck out like sore thumbs.

Perusing the downtown area, I did manage to find some additional souvenirs, but my heart was no longer in it, as the ship model I had prized, and all the other previous treasures, were now ashes, without any hope of replacement. But, even with the destruction, there was still much to see and evaluate, as I was in a strange and mysterious country with many extraordinary sites to see, and a lot of what was left to try to assess. While much of old Japan had been destroyed, what was left of the places and people who were still there, contributed to an experience that could never be duplicated or forgotten.

## Saving Prisoners of War

AS WE WERE SETTLING into the rigors of the Occupation, the remaining American prisoners of war were being repatriated, that is, those who survived the horrors of the Japanese prisoner-of-war system. These former prisoners were human skeletons, most of them barely alive from their brutal treatment. But, while they were being received from the prison camps and sent off to the United States, the stories of their hideous treatment, and photographs of their wasted bodies, became indelibly engraved in our minds.

There were many thousands European, American and Asiatic prisoners of the Japanese; and their treatment was so harsh that over one third of them died while in captivity, with many others permanently crippled from their harsh treatment. In contrast, it's estimated that only about one percent of American prisoners died in German captivity.

Surrender was not part of the Japanese military code. When backed into a situation from which there could be no honorable exit, the only course for the Japanese military was suicide. Because they had not committed suicide, prisoners of war were considered a low form of life, to be treated as such by Japan, with prisoners beaten, starved and executed as a standard practice. The radical officers in charge of the Japanese government completely rejected the Geneva Convention; and the brutal treatment of their prisoners was in direct violation of international law, which the Japanese officers' clique had completely rejected.

While surrender was forbidden in the Japanese Samurai culture, this was not the case with the British military, as shown by what happened in Singapore. There, a large British army of 85,000, a well fed and well-equipped army, surrendered to a small Japanese detachment of 30,000 soldiers, who were quite exhausted from fighting their way down the Malay peninsula, and were almost out of ammunition and other supplies. The Japanese commander could not believe his good fortune. As his demand for the British to surrender was a bluff, he was afraid that they would discover his numerical weakness and lack of supplies, and force him into street fighting, but they did not.

For the British prisoners of this abject surrender, it was a vicious arrangement, as many thousands of British prisoners died under conditions of forced labor, exhaustion and semi-starvation. Many of them, slave laborers, were forced to work on the construction of the Japanese "Death Railroad" in Burma and were forced to build the "bridge over the River Kwai." Contrary to the cinema version, the bridge was never destroyed, but still caused many British deaths and injuries while being constructed, and ever since, has been an attraction for Japanese tourists.

After the re-conquest of the Philippines from the Japanese military by the United States, an ominous general order was found in the Japanese archives. This order, to all unit commanders, stated that when the first American troops set foot as invaders on Japanese soil, all prisoners of war, 146,000 of them, were to be immediately executed. As we had broken the Japanese military code, we knew that this was no idle threat, and that while peace was being negotiated, their radical officers had begun executions.

On the island of Palawan, 150 American prisoners were herded into crude bomb shelters, which were then set on fire, and any survivors machine-gunned. On Honshu, several groups of captured American pilots were tortured before being slashed to death by samurai swords, their bodies buried in haste to keep word of these executions secret.

To ensure the safety of the rest of the prisoners while negotiations for the surrender were in progress, American rescue teams were parachuted to areas adjacent to Japanese prison camps, wherever they were located. The mission of these trained negotiators was to convince prison operators that the Emperor had decreed the end of the war; and with the Emperor's word in place, the execution of prisoners of war was no longer allowed.

Saving the Allied prisoners of war was carried out in a methodical way, with no mishaps. As soon as they were rescued, they were shipped for recuperation to American hospitals and other places of refuge. These remarkable rescue missions were accomplished without serious problems and were like many other missions accomplished by the United States during the war. These successful projects could take place only because of that incredible first shot in the war with Japan - the shot that allowed the United States to rescue the Japanese code from a sunken submarine. That rescue changed the course of the war and ensured that the peace process would continue as planned.

## Hunger in Japan

THE NEXT DAY, WE moved on to Osaka, the second largest city in Japan, and the headquarters for our portion of the Occupation. The city, a manufacturing center for various war industries, had been heavily bombed; the city was in shambles, food was running out, and the civilian population was becoming desperate. The 98th was assigned to a barracks in a former military enclave just outside the city limits. We hardly had time to get settled in, learn how to deal with the primitive plumbing (and other quaint Japanese technical aspects), when we were assigned crucial objectives to attend to: particularly, the relief of the hunger problem. Along with that, we were involved with the collection and destruction of enemy weapons and war materiel.

The need to relieve the Japanese food shortage was imperative, as over five million Japanese civilians were returning from that nation's far-flung conquests empty-handed and hungry. Because of the loss of food supplies from former colonies and, closer to home, the near destruction of the fishing fleet, the Japanese food shortage was acute. However, the Japanese people were trying to reverse this situation: in addition to growing much more rice than the usual amounts that had been grown locally, tangerines had ripened and were available at sidewalk produce stands, along with locally grown vegetables. And, in bombed-out areas, small, carefully tilled garden plots were flourishing amid the destruction. While these homegrown supplies were nowhere near enough, the hunger problem was eventually resolved with the arrival of significant supplies of food donated by the American government.

The near starvation of the Japanese people in the later days of the war became an issue with some Americans. In their view, the atom bomb was not needed, as we could have waited for starvation to force the Japanese to surrender. However, Joseph Grew, for many years the American ambassador to Japan, and an acute observer of the cultural mores of the country, did not expect Japan to ever collapse. In his book, *Ten Years in Japan*, he stated, "I know the Japanese people. I lived there for many years, and I know the Japanese people intimately. The Japanese will not crack. They will not crack morally or economically, even when eventual defeat stares them in the face. They will pull in their belts another notch, reduce their rations from a bowl to half a bowl, and fight to the bitter end."

Ambassador Grew was probably correct, as far as it concerned the civilian population. But because of a factor known only to a few, this discussion is probably somewhat moot, concerning the needs of the bureaucracy and the military. This is because of what I learned when I was part of a military guard duty group who, along with financial experts, were assigned to oversee the financial procedures of the Imperial Mint that related to the Occupation.

While much of the city of Osaka had been destroyed, the area where the mint was situated was unharmed, and along with the rest of my platoon, I was sent there. We were charged with the task of ensuring that the American financial experts went about their work at the mint without interference.

The mint, with an art deco style facade, would have blended in well with the buildings in any European or American city. It was an imposing sight: the marble facade was graced with the Imperial Chrysanthemum, a giant bas-relief in gleaming gold. As I stood guard outside the entrance, I could not help but be awed by the building and its impressive symbol of Imperial Japan. The workers at the mint, mostly women, evidently were also impressed by the majesty of the building, as they bowed to this Imperial symbol when entering or leaving the building. But, they also bowed to us, which made me somewhat uneasy, as I was not used to being bowed to. However, it was pleasant to see so many attractive young ladies, even though the language problem made communication with them somewhat difficult.

While the buildings comprising the mint hovered majestically over the mostly burnt-out city, there was something strange about a fenced-off section of the grounds. This was an area guarded by sentries and enclosing many large stacks of some kind of metal. We later learned that these stacks of metal that were piled outdoors, and protected by barbed wire, were silver ingots. This very valuable precious metal was now being kept outside, because, in this time of semi-starvation, the vaults were filled with a much more important substance: bags of that staple of the Japanese diet: rice. While civilians were going hungry, this vital food supply had been hidden away for exclusive use by the Japanese military and the bureaucracy. I later learned that the hoarding of this basic foodstuff took place not just in Osaka, but everywhere in Imperial Japan where the military was stationed.

Because rice, their fundamental sustenance, was available to them, the five million-man Japanese army was not only well armed, it was also well fed. And, as these well fed and highly trained soldiers were willing to die for their godlike emperor, they would have exacted a heavy price on those of us who would have invaded their sacred homeland.

## Benjo, the Very Strange Japanese Toilet

IN ANY EXCURSION TO a foreign country, the most crucial information needed when you arrive there concerns the location of the toilet facilities. In Japanese, that phrase was "benjo, wa doka deska," the toilet facilities being the "benjo" and the rest of the question asks where the toilet facilities are located. But 70-odd years ago, when I was part of the post-war army of Occupation, most of the Japanese toilet facilities, when found, were very different, in that they were uni-sex.

In contrast to Western customs, the benjo was used interchange-ably by both males and females, with no separation by sex. It took some adjusting for me to get used to the lack of privacy in the benjos, but there was no choice. However, before the war, the few hotels that catered to Western visitors did offer a choice. They had separate men's and ladies' rooms as in the West; but to reassure the locals, a uni-sex benjo was also available.

Not only were the customs different, but the actual toilet mechanism used in the benjos was radically different. The Japanese toilets showed an acceptance of Western concepts of convenience and sanitation, in that they featured an apparatus that allowed for flushing. However, they were radically different from the toilets of the West, in that they did not have a seating arrangement. You just squatted over a ground-level flushing apparatus as you used it. While this imparted a considerable level of discomfort for Western users, squatting was not an unusual practice in the Imperial Kingdom. The Japanese were used to squatting, and often did, for example, while waiting for transportation at bus and train stops; or, if they had to wait for any reason at all, the average person would just squat. And, as they were used to squatting, it imparted no discomfort to the Japanese when using the benjo.

For me, the lack of bathroom privacy was off-putting, but not impossible. However, never having used squatting as a way of waiting, I never really got used to the Japanese toilets. Fortunately, the latrines constructed in our barracks after we moved in provided some comfort. While Western-style toilets were not available, the newly built army latrine facilities, constructed from wood, reverted to outhouse-style construction, which allowed you to sit as you used them. And, you just made sure you used them before leaving the army area; otherwise, if a toilet was needed, you would have to squat.

Back before the war, very few Japanese homes had flush toilets. Chamber pots were used to collect human waste, which was not considered waste, as it was then used. Collected evenings by "honey wagons" drawn by oxen, it was transported to farm areas for use as fertilizer, as it had been used for centuries.

As part of our programs of cooperation with the Japanese, the product of the army latrines was collected by the Japanese, minus the toilet paper, which had to be deposited in separate containers. The army waste thus became useful, as was the army garbage, which was also hauled away for some other use. Anyway, the old cliché, "waste not, want not," was not wasted on the Japanese.

# Cabarets (Night Clubs)

The Occupation's first task had been to fill a drastic need, which was to obtain relief from starvation for the Japanese people. Although remedying the shortage was the number one priority for the United States, it would take some time before hunger would be completely mitigated; and shortages still persisted for a while. Because of the food shortage, the Occupational troops were forbidden to eat at Japanese restaurants and had to subsist for a while on the uninspired chow from the military mess halls. Fortunately, the need for this restriction did not last very long. As the American G.I.s were adept at finding and exploring the various pleasure palaces in a short time, cabarets (nightclubs), catering to the Occupational forces, had popped up throughout city areas. Very popular with the G.I.s and other military personnel, the colorful and romantic atmosphere of the cabarets might have been described as a kind of "Oriental Arabian Nights". They featured attractive young ladies in colorful costumes serving alcoholic beverages, who were also available as dance partners. Drinks and dance partners were paid for by tickets purchased at the door of the cabarets, as no cash payments were allowed once in the cabaret.

At the beginning of the cabaret club phenomenon, the elaborate kimonos of the dance hall hostesses made jitterbugging, the rather strenuous American dance craze, difficult for the dance hall gals. But, it was not long before these elaborate costumes were modified and simplified to allow freedom to move with the dance music, such as the music was. During the years leading up to the war, and during the war, anything Western was banned, including popular music. And, as American popular music was as new to the dance bands as jitterbugging was to the dance hall girls, the results were somewhat irregular, and catch-as-catch-can. But the bands kept on trying, and the girls kept on serving drinks, collecting tickets, and dancing with the G.I.s. While the bombed-out city presented a somewhat dismal ambiance, a spirit of festivity prevailed in the isolation of the cabarets.

## Bob Wulfhorst's Kamikaze Interlude

AN UNORTHODOX WEAPON, THESE hidden airplanes - envisioned by the Japanese as their major defense against the American invasion - were tentatively tried out for effective use during the battle for Okinawa. The Japanese had created the death-dealing concept of Kamikaze warfare by attaching powerful impact bombs to each plane. Fanatical Japanese pilots flew these planes - effectively guided missiles - 350 miles from their home base on Honshu, ending these individual flights in suicide, by deliberately crashing their planes into American cruisers, destroyers and other warships anchored off Okinawa.

A good friend of mine who was there, Robert Wulfhorst, a young Navy lieutenant at the time, was the navigator for his ship, an LCS – a type of landing craft that was the smallest seagoing warship in the U.S. Navy. His personal recollection of the mayhem and chaos at Okinawa was a vivid reminder of the danger that the Kamikaze system activated there, and the enormous number of casualties that would have occurred if the war had not ended when it did.

During one of the epic battles there, while stationed on his ship, Bob was standing next to the anti-aircraft gunner when he saw a Kamikaze zooming in from behind. But the gunner, facing in the other direction, and because of the volume of sound, had not heard the fast approaching suicide plane. Bob quickly knocked on the gunners helmet to alert him, and - just in time - the gunner swiveled around and began firing at the approaching Kamikaze. Perhaps because of that, the enemy plane veered off, and, narrowly missing Bob's ship, crashed into the sea. It was a close call - so close that Bob could see the pilot's face as the suicide plane flashed by. After crashing, parts of the Kamikaze's fuselage floated up, along with the pilot's body. His remains were retrieved, and in the pockets of his uniform were photos of his wife and children. Knowing that he would not return to his family, the pilot had fulfilled his military and civic obligations.

# The Saving of the Shrine City: Kyoto

WHILE IN OSAKA, THE lack of transportation caused by the devastation of war made it difficult to visit many of the interesting sights in the area. The subway in Osaka, because it was underground, still functioned, and did go off to some fairly distant parts of the city. But, as most of the surface transportation was in ruins, the subways were always crowded well beyond capacity, which made traveling anywhere not very pleasant.

Taking off to go anywhere was usually a kind of a spur-of-the-moment happening. Because we were part of the creation of the newly organized Occupation, no one knew exactly what was occurring, or had occurred, and so we didn't really have any scheduled time off (or time on). And, as there were so many tasks to perform, if we wanted to get away, we just waited until there was a break in the routine, and then we just took off; which I occasionally did.

The place I really wanted to see most was Kyoto, the former capital, and the shrine city of old Japan. This city was a collection of antique wooden buildings that the very astute American Secretary of War, Henry Stimson, had saved from destruction. In charge of the armed forces of the United States, Stimson, a highly intelligent, knowledgeable leader, helped build the very unprepared U.S. Army into a superior armed force. Even with all of the work required in the creation of America's armies and navies, Stimson never lost sight of the importance of the preservation of the arts, including historic sites. This included authorizing the "monument men of Europe," a group of experts who rescued an incredible number of works of art that had been stolen by the Nazis.

There were few museums in Japan, as most art was privately owned; so there was no need to have monument men available to retrieve stolen art work. In the Pacific area, however, instead of attempting the rescue of individual works of art, an entire city, Kyoto, was rescued. This collection of irreplaceable architectural treasures was saved by Secretary Stimson. He ordered that not only these many-years-old architectural monuments be off limits to the war, but that the entire city be preserved intact.

As an infantry soldier in the Pacific theater, I had no part in the rescue of Kyoto; but, early on, I did get to see the result of this very major act of cultural preservation. For many centuries, Kyoto was where the Emperor, while worshiped as a god (but having little or no power), languished amid surroundings of beauty and luxury. Once a year, the Shogun, the real ruler of Japan, would visit the Emperor to pay homage to the concept of Imperial rule. But, in 1854, after the peaceful opening of Japan by a U.S. Navy detachment, the Shogun lost face, and eventually, command of the government. His power was transferred to a representative form of government, with the Emperor as spiritual and temporal leader of this new political entity. To enforce the change, the Emperor moved to Edo, formerly the Shogun's capital, and it then became Tokyo, the capital of Imperial Japan.

Kyoto, though losing political power (as it was no longer the site of the Imperial court ), retained its spiritual status, as it was the shrine city of Japan, with thousands of visitors arriving every season. During World War II, though, that fame put the city in peril. Kyoto, the prime symbol of old Japan, and a city of antique and very flammable wooden buildings, was on a list that put it in danger of annihilation from American incendiary bombs. Fortunately, this threat was never carried out, as Secretary Stimson personally made Kyoto off limits to any American military aggression.

In September, 1945, my outfit, the 389th Infantry, one of the first outfits in the Occupation of Japan, was quartered in a former cavalry barracks in Osaka. In this second largest city in Japan, the city center had been mostly wiped out by fire bombings. A few modern skyscrapers, evidently fireproof, sticking out amid the debris of the bombed-out city, gave us an idea of what had happened all over Japan, and what could have happened to Kyoto.

As the Occupation got under way, the American authorities assigned former Japanese military/industrial sites as targets to be occupied. My group's target was the Imperial Mint, where we worked many days for long hours. Eventually, with some time off, I decided to use this as an opportunity to visit Kyoto, the famous shrine city whose beauty I had heard of as an art student. While word was that the American government had put Kyoto on a "do-not-bomb" register, it was, unfortunately, not on the list of approved off-duty destinations for soldiers from the 98th Division.

However, I was afraid that if I did not take advantage of this particular day off, I might not get another chance. In the chaos of the newly-in-place Occupation, I figured that I could get to Kyoto without anyone noticing my absence; so, I just took off. The only problem was how to get there on my own.

While the subways were still operating, they did not function outside of urban Osaka; and, as gasoline was in short supply, surface vehicles were scarce. Fortunately, the trains still ran, and so I made my way to the station, seeking out the train that ran to Kyoto. The train was rather ramshackle, with coaches jammed with passengers, but I had learned beforehand that I could squeeze in with the engineer, paying my way with the common currency of the Occupation, cigarettes. As Japanese civilians had been without tobacco during the war years, the engineer did not object to my crowding into his compartment. He quickly pocketed the cigarettes, and the train got under way.

I had conveyed to the engineer that my destination was Kyoto; and when we arrived there, it was like an "open sesame" to fabulous treasures. Temples, enriched with multiple Buddhas, were nestled next to ornate palaces and colorful shrines, along with picturesque pagodas; all in park-like settings, with what I presumed to be Zen gardens tucked in among the temples. Just about everywhere I looked, in this almost deserted city, there was a panorama of architectural delight, complemented by luxurious greenery. I walked around in a kind of daze.

In spite of the many temples, the city had an almost European look, with wide boulevards lined with Poplar trees. Modern storefronts and office buildings, that were seemingly empty of people, gave Kyoto the flavor of an almost ethereal city. A small, antique-looking trolley car, bouncing along, completed the picture. This shrine city of old Japan, with about 1 ½ million people, had been the home of the emperors for the thousand or so years, when the Shoguns ruled Japan. When the Shogun was deposed, the Emperor moved to the Shogun's headquarters city, which up to then had been called Edo. At this point in time it became Tokyo, and the center of the Imperial Japanese government.

Just walking around the former capital, I found much to admire, particularly the large and beautiful temples. Some were fairly new, and others were quite old. At one of the older ones, the gate was huge, with bronze doors thirty feet high, decorated with griffons and dragons. Inside the temple, the chant of the monks provided an aura of transcendence for the great god, Buddha, whose statue inside sat in golden splendor.

Visiting Kyoto had been a magical experience that I will always remember. Even the walk back to the station, on streets bordered with graceful Linden trees, became part of the fantasy conveyed by Kyoto. It was a beautiful place, but seemingly without much human activity. Among my postwar experiences, my visit to Kyoto rated highly, and I would have liked to have stayed longer, but as my visit was unofficial, I had to get back to my outfit before the evening mess call.

Upon my return to the station area, the quiet continued; but, seeing the glow of electric light in a small store, I decided to investigate. Most of the objects displayed inside were similar to the pre-war junk that was Japan's main export at that not-so-long-ago time. The proprietor, a wizened, elderly hunchback did not seem to be happy to see me in his store, but a small offering of cigarettes made him a friend. Disdaining the trash displayed, I conveyed to him that I was interested in Japanese art. Acknowledging my interest, he went into a back room, bringing out some small, very beautiful Japanese prints; just what I was looking for. I selected a few; but, as my train was pulling into the station, I had no time to palaver; so, I paid him his asking price without attempting to bargain, and hurried to the station, and to my seat in the engineer's compartment. Taking the dirty, noisy train back to Osaka was a return to reality. And, as I rejoined my outfit in time for evening chow, I was pleased to find that I had not been missed, and that my unofficial expedition to Japan's shrine city was a memory I could treasure without qualms. All this, along with the examples of Japanese art that I had acquired, gave this small side trip a memory that lingers.

This classical city, along with its ethereal ambiance, was not known just to me; but, while this ethereal abode is known to many, the details of how the saving of Kyoto took place are not that well known. And there does not seem to be any official acknowledgement of the importance of the preservation of this historical rarity, nor proper recognition of the accomplishments of Secretary of War, Henry Stimson, who, among many other achievements, made the saving of Kyoto possible.

## Visiting a French Priest in Occupied Japan

AT THE END OF World War II, an endless panorama of fascinating sights and sounds accompanied Occupational duty in Japan. So, when my Army friend, fellow soldier, and fellow Rhode Islander, George Roy, asked if I would like to join him on a visit to a French missionary priest who lived somewhat away from Osaka (the city where our local American Occupational forces were stationed) I was flattered by the invitation, and delighted to join him, even though my French was rudimentary. George needed no help with the language, as he was not only of French Canadian descent, but was a native speaker of the language. George had asked me to join him not for language assistance, but because he did not want to go by himself on an unusual expedition to a strange destination.

When talking to George, I noticed that he always hesitated before answering; asked why the hesitation, he replied that he grew up in a French language-oriented, New England mill town, Woonsocket, R.I. As a result, George spoke only French at home, in his neighborhood, and in the parochial school he attended; even the sermons in his church were in French. George did not learn English as a small child, as there was no point in learning a second language when everyone he knew spoke French.

However, when he was about 12, as his parochial school education did not go beyond grade school, he was sent to a public school. There, he began to learn English, and spoke it through high school, although not at home. The reason he hesitated before answering was that he still thought in French and had to translate English into his French language-thinking capacity before replying.

Our Catholic chaplain (who was not French-speaking), knowing of George's language ability, had suggested that he might like to visit the missionary priest. Isolated in rural Japan, the priest had not had a conversation in French for many years, even before the war. According to the chaplain, the elderly priest would be delighted to have a visit from someone who spoke French, even George's slightly different Canadian French.

George and I discussed the prospects of a visit away from Osaka and agreed that it was an interesting idea. So, on our first day off from Occupational duties, following directions provided by the chaplain, and bearing gifts of packaged Army rations and several cartons of cigarettes, we took a train that stopped at the village where the priest lived. On arriving, among the debarking passengers were several men who, judging from their tattered uniform-like clothing, backpacks, arm bands and split-toed shoes, were perhaps ex-soldiers returning home from some distant military outpost. If so, there was no one there to meet them.

It was a quiet, not particularly impressive village. The largest building, the train station, was surrounded by an irregular cluster of small shops. The rest of the town, unpretentious two-story, tile-roofed wooden buildings, straggled off in different directions. In spite of a light rain, and again following the directions given to us by the chaplain, we were able to wend our way through twisting alleys, eventually finding the small church and the parish house of the missionary priest. We were received there by the priest's housekeeper, a somberly dressed, middle-aged Japanese woman. After consulting with the priest, she indicated, mostly through gestures, that the Father was busy, but that he would be happy to see us. In a short time, a rather beautiful young lady in a colorful kimono exited the priest's study, and we were ushered in.

The priest, an elderly man - thin, almost gaunt, and walking with a slight limp - was dressed simply, in a plain black robe and black skull cap. He was delighted to see us, and to talk again in French. But, if his eighty-or-so years showed in his physical frailty, his calm face and contemplative eyes revealed a quiet dignity, and an almost mystical presence.

He introduced himself as Pere Armand, and we followed him into his study, where he motioned for us to be seated. The small, faded, old-appearing study contained few objects. The usual crucifix hung on a wall opposite a picture of the Virgin Mary. In back of where George and I were seated, a copy of Millet's Angelus was ensconced. Two large bookcases, filled with French and Japanese books, occupied another wall; and we sat around a small table covered with a cloth, so old and faded, the pattern could scarcely be determined. On the table, a well-worn bible rested.

The conversation began with a discussion - with George translating - about the priest's last visitor, a young lady whose father, a longtime parishioner, had died. Although she had fallen away from her father's beliefs, she wanted the celebration of his departure from this earth to be in accordance with his beliefs and had been there to make arrangements for the ceremony. During our conversation, Father Armand cradled in his hand, and occasionally puffed on, a kind of Japanese pipe. This common device (a mouthpiece connected by a long reed to a small pipe bowl) which held an upright hand-rolled kind of cigarette, rarely left his hand, even though the ersatz Japanese tobacco could not have been a very satisfying substitute for the real thing.

He was pleased with the gifts we'd brought, particularly the cigarettes.

"Merci bien," he said, putting aside the pipe. After lighting an American cigarette, he proceeded to smoke it down to almost the end, using a kind of home-made roach clip. He then lit a second cigarette from the first. Real tobacco was an unaccustomed luxury, and he relished the pleasure that smoking brought him. As Vichy France was part of the German, Italian, and Japanese Axis, he was spared internment in a Japanese prison camp, places where few survived. But, life had not been easy during the war years, and like the Japanese people, he had been deprived of many things, including tobacco and meat, with food in general being very scarce.

As Pere Armand smoked, he chatted away about his life, almost as if he were talking to himself. A most important, almost priceless, thing he missed, he told us, was conversation with fellow Europeans, and also his newspapers from France. Although slightly out of date by the time he received them, they had provided a lifeline to the world. Since the war, the newspapers no longer came, but perhaps now, he might be able to renew his subscriptions. This and more he related to George in a kind of meditative way. It was too much for George to translate word for word, but I was kept generally informed by an occasional summation from my friend.

Pere Armand continued to talk in his high soft voice, and I was somewhat able to follow the conversation. Obviously, the good Father was enjoying the luxury of speaking again in his native language. But, from time to time, George would be hesitant in his replies. His French-Canadian dialect - old French that included some English words - differed from European French. And so, some words, as spoken by George, were unintelligible to Father Armand; and in the same way, the Priest's words were often strange to George.

Occasionally, because of the language difference, they had to stop to make a word-search adjustment before their conversation continued. For the most part, I did not try to enter the conversation. But when George was trying to describe the new ways with food, he became stuck when it came to dehydrated potatoes. His problem was not just in explaining dehydration, but because he was using the French Canadian word for potato, *patate*. Here I came to the rescue with my high school French: "*Pomme de terre!*" I exclaimed, and the new food was made understandable.

In addition to the cigarettes, we had brought some packaged Army rations for the Father; and on a previous visit, the Army chaplain had provided Spam. An American staple, basic to army chow, this canned meat was so common to us that it was at the point of becoming a cliché. But Spam was the first meat the Father had eaten in four years, and "*tears came into my eyes,*" he said, "*when I tasted it.*" At this point, Pere Armand showed us how loosely his robe hung on him. His face, as well as his body, was thin, his cheekbones protruding under his parchment-like skin; but he had an aura about him of peace and tranquility that was palpable and unforgettable.

We eventually left Father Armand to his lonely, but I think satisfying, existence, saying our *au revoirs*, and walking out into the light rain, eventually finding cover at the station. When our train came, as I found a seat, I glanced out of the window to the other side of the station and saw, standing there under an umbrella, the beautiful lady who had been consulting with Father Armand. The sight of this colorful vision, waiting in the rain, became a memorable footnote to a remarkable event.

## Sayonara, Japan

THE WHOLE OCCUPATIONAL EXPERIENCE was completely different from what we had trained for, and yet it was such a significant success that the results were, in a way, mind-boggling. This came about when our intensive assault training suddenly stopped, and we went from our concentration on death-dealing maneuvers to giving generous help to an almost prostrate former enemy. Instead of forcing financial reparations on an evil empire, the United States instead donated tons of food to an almost starving population.

In the beginning of the Occupation the challenge of completing the massive military disarmament of Japan and, at the same time, relieving the hunger crisis (combined with the visual and physical aspects of a strange and alien culture) made it an almost overwhelming experience. While we could feel much pride in the American system and our accomplishment, military and political, there were still questions about the Japanese psyche.

Germany has recognized, and made amends for, the monstrosities of its Nazi past. But in Japan, while there is what the Washington Post called "a (very) short history of Japan's war apologies," Japan's prime minister, Shinzo Abe, has stated that while he has regrets, he declines to apologize for Japan's WWII actions. What were called by the allies "class A war criminals," were, Abe stated, " not criminals under the laws of Japan." While some of those most guilty of heinous crimes were executed, many others, equally guilty, continued their association with the government and industry of Japan; and in general, there is little remorse shown for war crimes. Among the favorite destinations for Japanese tourists are Pearl Harbor and the bridge over the River Kwai (which, in contrast to the movie version, is still standing). These two favorite destinations have in common the fact that they were the sites of the only Japanese victories over the United States in WWII. In the 1923 earthquake in Tokyo, the very beautiful Imperial Hotel, a noted edifice designed by American architect Frank Lloyd Wright, was the only modern building in Tokyo to survive, and after the war was the headquarters for the American Occupation. Shortly after the Occupation was considered to have been successful, and the Americans military went home, the Japanese removed this symbol of American success. They tore it down.

But there is hope. The fully independent Japanese government and the economy continue to flourish. Perhaps the hope exists because the American Occupation showed that democracies could not only win battles but could also govern civilian populations. For us occupiers, the visual and physical aspects of a strange and alien culture was an almost overwhelming experience. However, in spite of the mental flip-flop necessary to go from fighting a war to creating a new democracy, everything worked out the way it had been planned. We did what we were supposed to do and did it well. When it came time to leave, we left without leaving major problems behind us. The trip home took place when we became eligible, through the army "point" system, to very thankfully wend our various ways to our homes.

The young recruits who took over Occupational duties when we left were about the age I was when I enlisted. However, they had not gained much military experience, as they had been posted on full-time guard duty since arriving in Japan. Even though the U.S. military offered financial inducements for any of us who would stay on to help with the training of the recruits, not many agreed to remain, and most of us began to leave as soon as our points allowed.

My homecoming facility was via a "Liberty Ship," which took the short way from Japan to the United States by way of the Arctic Circle. It was a calm, quiet trip, alive with the expectation of civilian life. But, while on board ship, unencumbered by any responsibilities, I could think back on what I had been doing for the past three-and-a-half years; and while I really disliked Army life, for me it was not all bad. Among the positive elements I considered were that I had traveled to, and was stationed in, interesting and exotic places, I had gained many friends, and had extraordinary experiences that would be with me long after leaving the Army. In spite of what had often been proclaimed officially, some of my special duty assignments had given me insight into events that otherwise would never have been disclosed. All in all, my experience while serving the Army of the United States was an incredible education that otherwise I would never have experienced.

# Other Recollections

## Chaos in the Philippines

MOVING FROM KAUAI TO Oahu was not just switching from jungle training to beach assault maneuvers; the emphasis had changed entirely. From this point on, the invasion of Japan was our total concern, and all of our activity was directed to that goal.

But, in addition to learning from the revised training, I also learned a lot from new friends, many of whom I met in the various military clubs and bars available to non-coms. And, from their off-hand and uncensored comments, I learned about the disasters that kept happening in the early days of the war in the Pacific. While I was not there at the time, these were eye-witness reports that seemed authentic, and were vivid descriptions of unlikely events. But, even so, it was hard to believe the chaos they described - first, at Pearl Harbor, and almost immediately after that, in the Philippines. I kept getting negative reports from several of the groups of non-coms that I hung out with; and while they differed in describing certain details, in general, they concurred with the total picture. While some of our conversations were idle chitchat that was of little importance, all of it was considered confidential, as much of it put the military in a bad light. In fact, a lot of it was really hard to believe, and in certain cases some of these events have never been fully explained. For example, the most difficult to believe of these accounts concerned the absolute chaos that took place in the Philippines. And even today, much of that fiasco has never been explained.

What we heard was that a day after the attack on Pearl Harbor, the Japanese invasion of the Philippines began. In spite of having been immediately informed of the assault on Pearl Harbor, American opposition to the landing was minimal and disorganized; and strangely enough, no American planes were in evidence while the Japanese troops were landing. Nor were any American planes, or much of any other opposition, in evidence as the Japanese advanced toward Manilla.

The first Japanese air attack in the Philippines took place shortly after noon on the second day after the Japanese army had landed. Following the complete destruction of the American naval base at Caviste, the Japanese air force then went on to Clark Field, the main base for the U.S. Air Corps in the Philippines. All of the American planes at Clark Field, including 35 B17 bombers, scout and observation planes, were destroyed while sitting on the ground. This occurred despite the rule that in combat zones, 50% of all American planes had to be in the air at all times.

The retired American officers, who had volunteered to come to the Philippines to train a fledgling Philippine army, were now faced with a military problem that was beyond the scope of their training or experience in the field. All of the American military positioning in Asia had been based on the powerful American fleet. A fallback position evidently had never even been considered; and so, with no plan to guide them, and with the fleet in a shambles, pandemonium ensued.

General MacArthur, like many of the balance of American officers, had always proposed aggressive action, and had never believed in defensive maneuvers. So, when the Japanese attack came, and with the fleet now demolished, there evidently was no defensive plan.

In the Philippines, the Japanese air force wreaked havoc with the American air potential, and no American planes survived. To this day, no one, including General MacArthur, had ever explained the lack of any credible defense plan for the Philippines, or why none of the American military planes stationed at Clark Field, or any other military base, ever got off the ground. The only explanation, at least as told by my pals, was that the whole thing was so screwed up that it was a fiasco that had no possible explanation. And even today, no rational explanation has ever been offered. And, although it seemed hardly possible, for a while, things got worse.

Fortunately for me, by the time I got to Oahu, everything had changed, and we were beginning to win the war. As the 98th Division was included in the lineup for the invasion of Japan, all of these reports, opinions and comments from those who had been in combat, and others in the know, were meaningful. And so, these comments and criticisms of what had happened on the battlefield were retained in my memory and were tucked in next to the "elephant in the parlor."

*Special Note: Perhaps because of the total destruction of the Army Air Corps at Pearl Harbor and in the Philippines, along with other incidents of the misuse of airpower in the war, an important change took place. The air wing of the United States was taken from the Army and made into a separate service. It became an independent entity, The United States Air Force.*

## MIDWAY: THE EPIC BATTLE OF THE PACIFIC WAR

THE USE OF SUBMARINES at Pearl Harbor by Japan, as part of their assault on the American naval base at Pearl Harbor on December 7, 1941, was a serious mistake. It allowed the United States to retrieve the Japanese code from one of their technically advanced submersibles that had been fired on and sunk while trying to enter Pearl Harbor, shortly before the arrival of the Japanese planes that destroyed the United States fleet on that day. As the U.S. had no knowledge of how these small two-man submarines worked, the remains of this one, discovered on the ocean bottom near the entrance to the harbor, was raised and minutely examined for its technical innovations.

While these initial examinations were important, it was in the scrutiny of the documents on board the submersible that an extraordinary find took place: the current Japanese naval code. This find was a revelation of extreme importance, as it set the stage for the turning of the tide of the war in the Pacific in favor of the United States.

While the U.S. Government had in the recent past broken the Japanese diplomatic code, up until this point, the Japanese naval code had retained its secrets. As opposed to the almost unbreakable German *Enigma* code that used mechanical encipherment, the Japanese naval codes were book ciphers that were comparatively easy to break, particularly since the U.S. Navy now had the code book from the sunken submarine. From that point on, because of our having broken the code, the U.S. usually knew the plans of the Japanese military, as they were broadcast in codes that the American military now understood.

Additionally, there was another lucky break of major military importance for the United States. While much of the United States Pacific fleet had been destroyed in the attack on Pearl Harbor, three American aircraft carriers, complete with planes and crews, had been spared. These critical assets had been involved in military maneuvers away from the Hawaiian Islands, and were immediately available for combat, a readiness that was soon put into action.

After the attack on Pearl Harbor, in rapid succession, much of what was left of China, the Philippines, Wake Island, Guam, Hong Kong, Burma, Singapore, the Dutch East Indies, Indochina, the Solomon Islands, and almost everything in between, had succumbed to the Japanese military. Now, however, the tide of the war began to turn: with the Japanese code broken and our aircraft carriers available, America was ready to take on the formidable Japanese Empire.

Only six months after the disaster at Pearl Harbor, the American aircraft carriers, with their skilled crews, excellent aircraft, and capable pilots (aided by the knowledge gained from breaking the code) gained the upper hand in the first decisive battle since the attack at Pearl Harbor, and inflicted on the Japanese their first defeat of the war. This all-important achievement began at Midway Island, an American base on a remote atoll a thousand miles from Hawaii and halfway to Tokyo.

Being able to decipher the Japanese code, the American navy knew that an invasion of Midway was imminent; and so it was decided to ambush the Japanese by luring them to attack this important American base, when our Navy wanted them to. By cable, from Hawaii, the Midway base was told to send out a false radio broadcast, without encryption, saying that the desalinization plant at Midway had broken down, leaving this barren island without potable water.

The Japanese, intercepting the false broadcast, and knowing that a lack of water would make life on the atoll impossible for the defenders, had all the stimulus they needed for an attack, as they had indeed been planning to assault Midway. As the Japanese navy drew closer to Midway, the American navy was quickly able to learn the Japanese battle plans and where their ships, particularly their aircraft carriers, would be located. With this information at hand, and while the Japanese planes from their carriers were futilely bombing the hunkered-down garrison at Midway, the U.S. Navy planes attacked the enemy carriers when they were most vulnerable, while their planes were bombing Midway. The ambush was successful: in this first all-aircraft-only battle, the U.S. Navy carrier teams managed to destroy four of the six Japanese carriers, with only one American carrier put out of action. The loss of their aircraft carriers and their planes, pilots and crews devastated the Japanese military capabilities. And so, Midway is considered the most important American victory of the Pacific war.

It was almost inconceivable that the American navy, with most of its Pacific fleet destroyed by the Japanese attack on Pearl Harbor on December 7, 1941, was able, by June 7th, only six months later, to contest and defeat the hitherto unstoppable naval war machine of the Japanese Empire. Incredibly, something that "could not happen," did. It is important to note that the significance of the battle of Midway went beyond the Pacific war, as it not only changed the tide of the war in the Pacific, it influenced the course of world history.

A bonus from breaking the Japanese code: the U.S. Navy's control of events allowed them to film an actual battle while it was in progress. A Navy-sponsored film crew, headed by the famous director, John Ford, in a protected position on the Midway atoll, was able to film the Japanese attack as it happened. The footage from the battle became the spectacular documentary, "The Battle of Midway." When shown, the raw power of this film caused a spontaneous reaction. It brought home to the American people the reality of war, and was, at the same time, a most positive reinforcement for the morale of the American public, as it assured them that the tide of the war had indeed turned.

The film producers of Hollywood, and those working directly for the military, produced many interesting training films. Because the quality of the film was so high and the message so important, "Midway" remained a favorite training film among those shown to us.

# The Submachine Gun

IN BEACH ASSAULT TRAINING, the main American weapon was the standard semi-automatic M-1 rifle; this was a somewhat cumbersome weapon, requiring a trigger squeeze for every shot fired. On the other hand, the submachine gun, an American invention, was a weapon entirely suited for combat in close quarters. Easy to use and completely automatic, it could spew out a hail of gunfire. In contrast to the eight-bullet clip of the M-1, the ammunition drum of the submarine gun could hold 100 bullets. In every way suited as a military assault weapon, the submachine gun was used in America by the FBI, and in combat by the Army of the Soviets.

As the Japanese shoreline defenses were known to be even more effective than those encountered in Normandy, our training for the invasion of Japan included practice in the use of Bangalore torpedoes, along with accelerated training in hand-to-hand combat and bayonet practice. But left out of consideration was potentially the most important weapon for combat in enclosed spaces (such as pillboxes and other types of fortresses that were in place to protect the Japanese homeland). That super weapon was the submachine gun.

# An American Flying Boat Lands in Japan

IN WARTIME, THERE IS always the hurry-up-and-wait effect, where spurts of activity are often followed by periods of no action. The only official suggestion for what to do in these intervals of breaks in the routine is, "smoke, if you got 'em," and, as cigarettes were part of every food ration, you always "got 'em." But sometimes, with these vacant time-offs hanging in the air, someone would begin a story, often about something that had taken place in the war. Quite often, they were fragmented bits, but sometimes a story got told all the way through. One of these that I remembered concerned a PBY (flying boat) wartime flight over Japan that had unusual results.

As the story was told, an American PBY officer about to be rotated back to the United States was required, by regulations, to fly with his replacement team on their first mission, without comment, as they went through their routine. If all went well, he could then turn over the command of the PBY to the new crew. All seemed to be okay on the test flight until the new navigator confessed that, while he thought they were over Okinawa, the land below looked strange, which was not reassuring, as they were almost out of gasoline.

The about-to-be-rotated officer took over, and, having spent considerable time flying over Japan, knew the lay of the land. In the ending days of the war, American planes ruled the skies without opposition, and many flights had been made over this area. By looking down and knowing the contours, the experienced officer saw that instead of Okinawa, they were over Japan, with the fuel about to run out.

As there seemed to be no way out of this potential disaster, he suggested that they bring the flying boat down to the Japanese bay below them. And, as there seemed to be no other option, they would just take their chances by surrendering to authorities. This meant that they would probably become inmates in a prisoner of war camp, even though the reputation of the Japanese prison camps was of sadistic treatment, particularly for captured pilots. And, of course, the furlough he had been so looking forward to was now out of the question.

They set the PBY down in the bay, and as twilight fell, they waited for something to happen. Nothing did, until a ferryboat crossing the bay came close enough for them to taxi over and intercept it. The ferry stopped at their waved command, and two of the PBY officers boarded it, thinking to start the process of turning their crew over to the authorities, wherever the ferry was going.

As they sat on the ferry, grimly anticipating their future, a young Japanese woman joined them. Speaking perfect English, she informed them that she had been educated at Berkeley. When they asked her if there was a naval base nearby, she pointed to some industrial buildings on the other side of the bay. As darkness was setting in, they were pondering what to do next, when to their mystification, she told them that they would like the little town they were headed to; and further, she told them, she would take them to the best inn, which she did. But, before leaving them there, she said something to the man who seemed to be the innkeeper.

Not knowing what to do next, they pantomimed their need for food and sleep. They received a simple meal and a futon to sleep on, along with blankets to keep them warm. In the morning, they were given another meal, but when they offered American money as payment for the lodging and meals, it was refused. Returning to the ferry dock, they found a ferry ready to go out, and again obtained a ride out to their flying boat, still mystified by the strange but welcome treatment.

However, if it was at all possible, they still needed some way to get back to their base in Okinawa. And, having learned from the young woman of a naval base directly across the bay, they used the PBY's remaining gasoline to taxi over to it.

Seeing what seemed to be a seaplane ramp, in what was obviously a naval base, they cruised over to that, expecting arrest by the military authorities. Instead, Japanese sailors came over, seemingly friendly, and when the Americans indicated the fuel problem, proceeded to refuel the PBY. When finished, the ground crew bowed and waved them off, obviously not expecting payment; and the nonplussed Americans flew on to Okinawa.

When they arrived back at their base, the only explanation anyone could give them for this very strange treatment was that, because peace talks were in progress (which the PBY crew had not known about), the Japanese evidently mistook them for part of the peace mission. And, as the God-like Emperor in a radio broadcast had said, "The war is over," for the Japanese people, it was over. And, as the adventure of the PBY crew had shown, without a doubt, the war was over.

# General Douglas MacArthur:
## The Shogun for a New Japan

WITH THE MILITARY OCCUPATION of Japan fully in place, the next step in the takeover by the United States was the governing of what was still, technically, Imperial Japan. To do this, the United States created an Occupation that was successful in every way, not only rescuing a former enemy from near starvation and chaotic ruin, but turning Japan from a military dictatorship into a vibrant democracy. From the beginning, the Occupation, led by General Douglas MacArthur, enacted well thought-out military, political, economic and social reforms.

MacArthur was a natural for this assignment, as he understood the way the Japanese thought process worked. He had lived in Japan when his father, General Arthur MacArthur, was military attaché to the U.S. Embassy in Tokyo. They were much alike, as father and son excelled in military combat, and both were awarded the Medal of Honor. Growing up in a military family, and eventually becoming a five-star general, Douglas MacArthur was the personification of a Shogun, a leadership figure historically envisioned by the Japanese. His always "on stage" aura of power, reinforced by the formidable farsighted thinking of the planners of the Occupation, created a workable structure for the Occupation.

The planning, based on policies of the American "New Deal," was the kind of thinking that led to the Marshall Plan. For the first time in history, a plan was put in place to help a formerly savage and implacable enemy, not just recover from the ravages of war, but to thrive. As soon as peace was declared, teams of United States financial and military experts fanned out all over Japan, working to break up the Samurai enclaves as well as the huge estates controlled by the Zaibatsu, the legendary power brokers of Imperial Japan. The purpose of all this political organization by General MacArthur and his staff was to encourage the development of democracy in Japan.

Much depended upon the relationship of the people of Japan to the Occupational forces, which were personified by General MacArthur. And, as the concept of MacArthur as a kind of Shogun grew in importance, there was a resurgence of interest in the historical Shoguns. For over four hundred years, Japan was ruled by Shoguns, a succession of military dictators who ran Japan with an iron hand; while the Emperor, a ruler without power, lived quietly in Kyoto, the historical capital. For centuries, an isolated island nation where nothing ever changed, Japan had been compelled by the Shoguns to renounce any contact with the outside world. However, in 1853, an American fleet of well-armed steamships arrived in Japan. Steam, being a technology unknown to the Japanese, showed that the Shogun lacked military control and thus, political power. He was forced out, and the Emperor, for the first time in centuries, gained control of his Empire.

However, the concept of a Shogun, a military dictatorship running Japan, still lurked in the Japanese psyche. MacArthur's dynamic physical presence and weighty directives ensured the acceptance of the carefully conceived American plan for peace. In essence, MacArthur had become the American Shogun, and with his staff, took over the challenge of running Japan. While some war criminals were punished, and a few were executed, no guilt was implied for the Emperor or any of his family, and the Imperial court became an important ingredient in the workings of the Occupation.

Our main mission was to monitor Japanese government activities. The American Occupational administration allowed the Japanese government to continue to operate as it always had, but under the supervision of the American advisory forces, and to continue working toward becoming a representative democracy. In effect, there was not an American military government, except in an advisory capacity. Our main push was to get the Japanese government and industries in working order and able to operate in a democratic manner, with some restrictions.

American programs put in place included land reform, which was initiated by the U.S. and then turned over to the Japanese for execution. Three million peasants acquired 5,000,000 acres of land from the Samurai and other major landholders, including the Zaibatsu, the giant industrial and financial conglomerate which had owned much of Japan since the early history of that country. Also, in a dramatic reversal of old-time restrictions, universal suffrage was enacted. For the first time in history, women gained the right to vote and to own property. In 1946, this resulted in a radical switch in politics, with one third of the votes being cast by women, and in the election of 39 female candidates.

The reorganization continued, and in a relatively short time, Japan was repaired, restored, and back in business, with friendly relations with America predominating. The incredible success of this farsighted military/political pacification encouraged the adoption of the Marshall

Plan in Europe, which created a postwar, fair and just European political entity. But the fair and just Occupation of Japan came first and showed the way to arrive at a lasting peace.

## Sex Life in Occupied Japan

GENERAL MACARTHUR IMPOSED STRICT rules of behavior for the troops of the Occupation, and for the most part, they were obeyed. Included was the imposition of severe punishment for rape.

The Japanese had always had prostitution districts in their cities. However, in Tokyo, that district was burnt to the ground in the fire bombings that destroyed much of that city. Immediately after the end of the war, new prostitution districts were created, opened, and made available to American servicemen. But, before Americans were allowed to make use of these brothels, it was required that a physical examination of the workers be made by the brothels' official doctor. The doctor, after having examined the inmates of the new accommodations, reported that there was not a serious problem.

However, when the brothels were opened for American servicemen, it was not long before an extremely high rate of venereal disease was reported, when the American troops who had used these facilities had their monthly physical. The Japanese brothel doctor who had originally examined the workers was queried, as he had reported that only a relatively small percentage of the workers were diseased. In questioning his methods, it was found that his physical examination of the sex workers for venereal disease had just been above the waistline.

Because of the high rate of venereal disease, General MacArthur eventually ordered the brothels off-limits for American servicemen. Immediately following that order, reports of rape rose sharply. However, with strict enforcement of regulations that followed, the number of sex crimes reported declined sharply, and were not common in Japan during the remainder of the Occupation.

# Dwellings in Imperial Japan

MY GRANDFATHER HAD DESCRIBED Japanese houses to me and John Roenig, who was part of an intelligence group I worked with, and provided information in greater detail. He had lived in one when he was part of a team of recent Annapolis graduates who, just before the war, had been selected to learn about Japanese culture by living in Japan in a Japanese house. He described in some detail how houses were constructed, and this information was confirmed by my own observations.

The basic construction of the outside walls of a Japanese house was of woven bamboo siding that was attached to corner posts; the siding was then covered by an adobe made from a mixture of rice-straw and mud. An overhanging roof, usually of ceramic tile, extended out over the walls, protecting the adobe sidewalls from the elements. While this does not sound like durable construction, this was not important, as Japanese homes were not built for permanent usage, but were replaced every twenty years or so.

The interior of the house was one room that varied in its usage. Sliding panels were used to create bedrooms, but in the morning, bedding was put in a large closet, furniture would be brought out, and the sliding panels that were used to change the size of the rooms were set aside. These sliding walls, when needed, were used to create other temporary living quarters. But, as the main room had several uses during the day, these panels were slid back against the walls, thereby creating a daily living space.

The kitchen, bathroom and toilet were separate attached rooms, with the cooking facilities provided by a small charcoal stove which also heated the house. Charcoal was in short supply during the postwar years; and as houses were not heated by anything other than these small cooking stoves, it often was chilly. So, in cooler weather, at least to provide more warmth, additional kimonos would be worn.

Back then, there were few flush toilets, and chamber pots were usually utilized. In the evening, the chamber pots were emptied into containers on the street, to be picked up in the morning by fertilizer wagons that took their contents off to local farms.

In the cities, flush toilets were sometimes available, but these were Japanese-style, with no seat. The toilet consisted of a ceramic opening in the floor, housing a flushing mechanism, and was utilized by squatting over them. While Western-style hotels and other modern buildings often had separate male and female lavatories, with Western-style toilets that one could sit on, many of them still maintained an additional facility for their local customers: the traditional Japanese, male/female co-ed toilet, known as the benjo.

## Religion in Imperial Japan

AFTER THE SHOGUN WAS deposed in the 1840's, religious freedom was granted in Japan, where the religions of the Western world had been banned for three centuries. But, the Christian missionaries, when they flocked to Japan, found, to their amazement, that there already were some Christians living in Japan. Known as 'hidden Christians,' they surfaced after religious freedom was announced, and were the remnants of a larger number of Christians, most of whom were massacred by the Tokugawa Shoguns in the 16th century. Christians were originally tolerated by the Shoguns; and they flourished in Japan after the Portuguese saint-to-be, Francis Xavier, converted a large number of the working-class Japanese to Christianity.

However, an important historical event that happened in the Philippines alerted the Shogun to a potential problem. There, the Spanish military followed the Spanish missionaries into that island country and conquered the Philippines for Spain. The Shogun, fearing that this would happen, banned Christianity. Following the lead of the Shogun, the Samurai armies then killed off any remaining Christians still in Japan after the ban. The only exceptions were several small groups of Christians who, risking torture and death, went underground to preserve their faith.

Because anything material could be used against them, these underground worshipers did not keep any evidence that could cause them to be exposed. Their version of Christian dogma, after centuries of word-of-mouth translations, had been modified into a different version of Christianity from that of Europe's Roman Catholic Church. The Catholic missionaries to Japan wanted the hidden Christians to give up their version of Catholicism and rejoin the Roman Catholic Church. But the hidden Christians, after having gone through centuries of preserving their beliefs, risking torture and death for these remembered dogmas, felt that their version should be preserved.

So now, in Japan, in addition to those who follow Shinto and Buddhist beliefs, there are Protestant Japanese Christians, Catholic Japanese Christians, and Hidden Japanese Christians; and they all (about 2.3% of the total population) seem to get along with each other. But strict religious doctrines in general, are not that important to the Japanese, who usually support various religious holidays, no matter which domination is being featured: Such as Christmas, which amounts to less than 2% of the Japanese people but is celebrated by most Japanese.

Until the Occupation took over, the official state religion was a government-controlled version of Shinto, which consisted of native beliefs merged with Chinese Buddhist. This official conglomeration was used to support the Japanese militarist policies that led to Japan's current military disaster. So, the Occupation, in addition to allowing everyone to vote and to have religious freedom, decreed that the Shinto religion could no longer be an official organ of the national Government, although it continued to be part of the life of many Japanese people.

Restored to power in 1868 after the Shogun system was revoked, the Emperors, in their role as a spiritual power in Japan, moved to Edo, the shogun's capital, now called Tokyo, where they ruled in conjunction with the military authorities. At that time, in this rule of an all-powerful God/king made the Emperor the spiritual leader of Japan. Although radical officers had taken over the government in recent years, the people of Japan still approved of and believed in the status of the Emperor as their spiritual leader. After the Occupation, even after the Shinto religion was removed from the government, the people of Japan continued to approve of and believe in the Emperor's spiritual power.

And that was important to us, because when he said the war was over, it was over. The Japanese acceptance of the Emperor's statement allowed the Occupation to be peaceful and productive, and therefore we could get home sooner. Although Emperor Hirohito subsequently renounced his divine status, his spiritual aura, as the symbol of modern Japan, continues through his descendants, although their political powers are in abeyance.

## The Japanese Zero

AN IMPORTANT TASK FOR us in the Occupation was the collection and dispersal of Japanese armaments, including the several thousand Kamikaze planes hidden in underground hangars. The majority of those were the once feared Mitsubishi Zero, which had recently been outclassed by newly created American pursuit planes.

When the war broke out, the Zero, a long-range, super maneuverable fighter plane, was so far superior to any American plane that the official advice given to American pilots was that when they encountered a Zero, they should run away. Fast and powerful, the Zero was known as a nearly invincible fighter plane. In an April 1942 battle over Ceylon, with well-trained English pilots flying the Spitfires that had held off the German Luftwaffe, 36 Zeros took on 60 British aircraft, shooting down 27 of them, with the loss of just one Zero. At the beginning of the war, with the U.S. fleet crippled at Pearl Harbor, Japan ruled the seas; and because of the fearsome, agile Zero, Japan also ruled the air.

The origin of the brilliantly designed and flawlessly manufactured Zero has some questions, the answers to which probably will never be known. But aviation experts noted that before the war, Howard Hughes, the eccentric inventor and movie producer, had created a plane that was in many ways a prototype of the Zero. Hughes offered the U.S. War Department this innovatively designed plane, but it was rejected. Shortly after that, Japan began to produce the Zero.

In a conversation with an acquaintance, whom I was told was an aeronautical expert, the possible Hughes-Zero connection was discussed. He had an interesting comment: he said that he was attending a conference at the Howard Hughes Aeronautical Hangar Museum, where Hughes's monster flying boat was kept. As he was being guided through the facility, he noticed a smallish plane off to one side. "What's that Japanese Zero doing here," he asked. "That's not a Zero," he was told, "that's the Hughes H-1, a plane that Howard Hughes designed."

When the war began, the Japanese Zero outclassed all of the American fighter planes; but as the war progressed, the United States greatly improved the quality of its armaments, particularly its military aircraft. The inferior American planes were replaced by the P-47, and particularly the P-38, which compiled impressive victories over the Zero. For example, an American ace, Major Richard Bong, flying a p-38, shot down forty Japanese aircraft in two years. While America innovated and reorganized its air force, the Japanese were unable to do so. Their remaining planes, seriously outclassed, were hidden away in underground hangars, to be used by suicide pilots as Kamikaze guided missiles against the troop ships of the upcoming invasion.

## Postscript

In June of 1950, the shaky postwar peace was shattered when North Korean troops, equipped and trained by the Soviet Union, began the physical aspects of the "cold war" by attacking unprepared South Korea. The young, untrained American soldiers who had been doing guard duty in Japan and Korea were thrown into the defense against the invasion. But, because of their lack of training, they were no match for the well-trained North Koreans; and in the initial conflict, many of the Americans were captured and committed to prisoner- of- war camps, just as had happened in the Philippines at the start of WWII. In many ways, it was almost as if the new so-called cold war was just a continuation of the epic war that had theoretically ended. It was just as that war, WWII, was, in a way, a continuation of the first world war, which had ironically been promoted as "The War To End All Wars."

General MacArthur recovered the initiative with a master stroke, surprising the North Korean military forces with a landing at Inchon - partway up the Korean coast. As there were not enough soldiers skilled in assault landings to make this attack, MacArthur used the always well trained United States Marines for this perilous landing; and the North Koreans, outflanked, retreated North.

A United Nations force, led by the United States, and created to evict the North Koreans, advanced rapidly up the Korean peninsula toward the border with China. American military intelligence now warned that a large Chinese force was moving toward Korea; but General MacArthur, who had little patience with military intelligence, took no precautions against the Chinese challenge, and continued the advance of his multi-national force. The enormous Chinese army then attacked and engulfed the spread-out and outnumbered American/United Nations forces, with many of them captured and sent off to prison camps. While this was going on, General MacArthur, who had always maintained that it was the duty of a commanding officer to be up front with his troops, had broken his own rule and had remained at the American Army headquarters in Japan.

As a result of this debacle, General MacArthur was removed from his command by President Truman. Upon returning to the United States, and seeking political support, General MacArthur made his famous farewell speech to the American Congress. He concluded it by quoting the old army song, *Old soldiers never die, they just fade away*. Eventually, however, having gained little political support, General MacArthur, now an old soldier himself, gradually faded away, retiring from the military and political limelight to a quiet, unobtrusive life as private citizen.

## About the Author

Having been on Special Detail during WWII, Richard Parker became aware of projects that were instrumental in aiding military action in the Pacific. Even before the war, Parker was attracted to that mysterious area since his family had connections to the Far East. Parker's grandfather had been a ship's surgeon for the Blue Funnel steamship line that sailed from Vancouver to the Far East. Conversations about that area were a constant focus at his house.

Parker grew up in Rhode Island where he attended Rhode Island School of Design (RISD), with emphasis on the art of the Orient. He was a student there when the war broke out.

After military service that included three and a half years in the Infantry, much of it in the Pacific area, Parker moved to New York to work as a freelance artist and promoter. Clients came from the worlds of fashion, professional football, the movies, textiles, railroads and manufacturers of military weapons.

Parker's promotions for the National Football League were so successful that Warner Brothers asked him to promote some of their major movies.

He did, and was invited to be their guest at the Cannes Film Festival celebration for Warner Brothers' 50th anniversary. At Warner Brothers he worked with Francis Ford Coppola on his first major movie.

Parker was the assistant to Coco Chanel in her fabulous return from self-imposed exile to worldwide fame. Says Parker, "The time I served as Mademoiselle's assistant was beyond comparison."

During the era of the Mad Men of the New York advertising world, Parker was an account executive for the giant advertising agency, BBD &O. There he promoted DuPont products before joining the international ad agency Bozell & Jacobs as a Vice President.

Taking early retirement, Parker moved back to Rhode Island to paint and eventually write. While his painting was well received, much of the writing that ensued was for Op/Ed pages in the prestigious *Providence Journal*.

Parker has authored two books including, "*The Improbable Return of Coco Chanel, As Witnessed by her Assistant, Richard Parker*," and finally *Pacific Memories: War and Peace in Far Away Places.* A third book about mayhem connected with art treasures of Africa; *The Benin Massacre*, is in progress.

# Richard Parker's Other Book

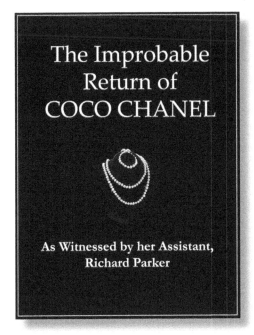

Richard Parker's recollections of his time as the assistant to the fashion industry icon, chronicles the untold challenges encountered in opening a new showroom for Chanel Perfumes in New York; the hand-to-hand corporate infighting between Gregory Thomas, the powerful Chairman of Chanel America, and Tom Lee, its legendary designer; and the ultimate resurrection of Coco Chanel's reputation and legend. Parker's insights and comfortable writing style bring this industry-defining event and its era to life in page-turning fashion.

He has also published articles in *Broadcasting Magazine* and *Our Town* magazine; three articles for the *South County Independent* newspaper, and ten articles for *The Providence Journal.*

Made in the USA
Las Vegas, NV
28 August 2021